AF522086

Florencia Montagnini
Editor

Environmental Services of Agroforestry Systems

Environmental Services of Agroforestry Systems has been co-published simultaneously as *Journal of Sustainable Forestry*, Volume 21, Number 1 2005.

Pre-publication
REVIEWS,
COMMENTARIES,
EVALUATIONS . . .

"TIMELY AND APPROPRIATE. . . . ALL THOSE INTERESTED IN AGROFORESTRY AND THE ONGOING ENVIRONMENTAL DEBATE WILL FIND THIS BOOK QUITE USEFUL. By presenting authoritative accounts of carbon sequestration scenarios in major agroforestry systems in the American tropics, such as silvopastoral and shaded perennial systems in comparison with pastoral and tree plantation systems, this book provides valuable insights into the emerging area of environmental services of agroforestry. A chapter on rural community perspectives on the issue adds the needed socioeconomic dimension to the discussion. Dr. R. Lal's masterful overview of carbon sequestration in natural and managed tropical forest ecosystems makes the book quite international in scope."

P. K. Ramachandran Nair, DrSc, PhD
Distinguished Professor
Agroforestry & International Forestry Director
Center for Subtropical Agroforestry
University of Florida

More pre-publication
REVIEWS, COMMENTARIES, EVALUATIONS . . .

"RECOMMENDED for scientists working on field-related topics, for instance on carbon stocks combined with agroforestry practices, but also to resource managers and decision-makers eager to find efficient and fair solutions through environmental services. Hands-on trials on carbon sequestration and biomass production are presented, explaining how research should be implemented in terms of complying with land tenure issues."

Dr. Henri-Félix Maître
Senior Research Officer
Forestry Department
CIRAD
Editor-in-Chief
Bois et Forêts des Tropiques

"AN IMPORTANT COMPILATION of quantifiable information related to environmental services with emphasis on experience gathered mainly on agroforestry systems in the neotropics. Individual chapters will be RELEVANT TO ADVANCED STUDENTS AND PROFESSIONALS involved in research and technology transfer in environmental studies, environmental services, biology, forestry, and particularly to those working in agroforestry systems in Mesoamerican countries."

Alfredo Alvarado, PhD
Professor and Researcher
Centro Investigaciones Agronómicas
Universidad de Costa Rica, San José

More pre-publication
REVIEWS, COMMENTARIES, EVALUATIONS . . .

"POLICYMAKERS, MANAGERS, AND ON-THE-GROUND PRACTITIONERS IN TEMPERATE REGIONS AS WELL AS IN THE TROPICS CAN BENEFIT GREATLY from the blueprints convincingly presented in this book. It brings together a body of literature from the tropics that explores the potential for incorporating trees in agricultural settings as a means of increasing carbon storage, both above and below ground."

Carl F. Jordan, PhD
Senior Ecologist
Institute of Ecology
University of Georgia

"ESSENTIAL READING for soil scientists, agronomists, and foresters. . . . This book will benefit any interested reader concerned with land-use planning, land management, agroforestry, or plantation forestry. . . . Very valuable, updated, and informative information on C-sequestration is included."

Gonzalo de las Salas, PhD
Professor of Ecosystems Analysis
Universidad Javeriana
Bogota, Colombia

More pre-publication
REVIEWS, COMMENTARIES, EVALUATIONS . . .

"A VALUABLE CONTRIBUTION TO THE LITERATURE . . . considers the benefits and barriers of two of the most popular secondary objectives in applying agroforestry as a management regime, carbon sequestration, and biodiversity conservation. . . . Particularly useful for development practitioners."

Kirsten Spainhower, MF
Development Specialist
Agriculture and Natural Resources Group
Development Alternatives Inc.

Food Products Press®
An Imprint of The Haworth Press, Inc.

Environmental Services of Agroforestry System

Editor

Florenica Montagnini

International Book Distributing Co.
(Publishing Division)

Published by

INTERNATIONAL BOOK DISTRIBUTING CO.
(Publishing Division)
Khushnuma Complex Basement
7, Meerabai Marg (Behind Jawahar Bhawan)
Lucknow 226 001 U.P. (INDIA)
Tel. : 91-522-2209542, 2209543, 2209544, 2209545
Fax : 0522-4045308
E-Mail : ibdco@airtelbroadband.in

First Indian Reprint 2007

ISBN 81-8189-189-9

Published under arrangement with Food Products Press®, an imprint of The Haworth Press, Inc., 10 Alice Street, Binghamton, NY 13904-1580

Printed at:
Salasar Imaging Systems
C-7/5, Lawrence Road Industrial Area
Delhi - 110 035
Tel. : 011-27185653, 9810064311

Environmental Services of Agroforestry Systems

CONTENTS

ABOUT THE EDITOR

Dr. Florencia Montagnini is a Professor of Tropical Forestry at Yale University School of Forestry and Environmental Studies. Her research focuses on variables controlling the sustainability of managed ecosystems (e.g., primary and secondary forests, plantations and agroforestry systems) in the tropics, with special emphasis on Latin America. Born in Argentina, she has a BS in Agronomy from the National University of Rosario, Argentina, and a Masters degree in Ecology from the University of Georgia, USA. She has studied and worked extensively throughout Latin America, and operates in collaboration with local institutions such as CATIE (Tropical Agriculture Research and Higher Education Center, Costa Rica). She is currently conducting research in Costa Rica, Panama, El Salvador, Guatemala, Nicaragua, Argentina and Mexico, on sustainable systems for restoration of degraded ecosystems.

Preface

This Special Issue of the *Journal of Sustainable Forestry* is one of several outcomes of the First World Congress in Agroforestry that took place in Orlando, Florida, USA, 27 June to 02 July, 2004. Several sessions and Symposia dealt with environmental services of agroforestry systems. Ten or even twenty years ago, when referring to the environmental benefits of agroforestry the main focus was on soil amelioration, control of erosion, microclimate control, alleviation of the effects of drought in semiarid areas, or other subjects that related to productivity or protection at a local or regional scale. But today, with increasing awareness of the globalization of human's effects on the environment–impacts on biodiversity, global warming–environmental services at a global scale (carbon sequestration, conservation of biodiversity) are an increasingly relevant concern in agroforestry. Proper design and management of agroforestry practices can make them effective carbon sinks, and they can also serve to protect biodiversity to different extents. Agroforestry can also have an indirect effect on carbon sequestration and on biodiversity when it helps decrease pressure on natural forests, which are the largest sink of terrestrial carbon and are also large reservoirs of biodiversity. Another indirect avenue of carbon sequestration is through the use of agroforestry technologies for soil conservation, which could enhance carbon storage in trees and soils. Agroforestry systems with perennial crops may be important carbon sinks, while intensively managed agroforestry systems with annual crops are more similar to conventional agriculture (Montagnini and Nair, 2004).

This Special Issue of the *Journal of Sustainable Forestry* includes a total of six articles that were presented at the First World Congress in Agroforestry. In "Soil Carbon Sequestration in Natural and Managed Tropical Forest Ecosystems," Rattan Lal presents several examples of

[Haworth co-indexing entry note]: "Preface." Montagnini, Florencia. Co-published simultaneously in *Journal of Sustainable Forestry* (Food Products Press, an imprint of The Haworth Press, Inc.) Vol. 21, No. 1, 2005, pp. xix-xxi; and: *Environmental Services of Agroforestry Systems* (ed: Florencia Montagnini) Food Products Press, an imprint of The Haworth Press, Inc., 2005, pp. xiii-xv. Single or multiple copies of this article are available for a fee from The Haworth Document Delivery Service [1-800-HAWORTH, 9:00 a.m. - 5:00 p.m. (EST). E-mail address: docdelivery@haworthpress.com].

Available online at http://www.haworthpress.com/web/JSF

carbon sequestration by agroforestry systems worldwide. Carbon concentrations in soils are not alone a good indicator of the capacity of a soil to sequester carbon from the atmosphere. When bulk density of the soil is used to calculate carbon pools in soils, sometimes the answer is very different as when only concentrations are taken into account. Carbon sequestration in soils is a "win-win" situation: those management practices that induce carbon sequestration at the same time help improve soil productivity and therefore are important for increasing food security.

In the next article, "Carbon Sequestration in Pastures, Silvo-Pastoral Systems and Forests in Four Regions of the American Tropics," María Cristina Amézquita and collaborators report on results generated by the "Carbon Sequestration Project, The Netherlands Cooperation CO-01002" on soil carbon stocks (SCS) for a range of pasture and silvo-pastoral systems prevalent in agro-ecosystems of Tropical America compared to native forest and degraded land. The results so far show that in tropical ecosystems, improved pasture and silvo-pastoral systems show comparable or even higher SCS than those from native forests, depending on climatic and environmental conditions (altitude, temperature, precipitation, topography and soil), and represent attractive alternatives as C-improved systems.

In our article "Environmental Services of Native Tree Plantations and Agroforestry Systems in Central America" (Montagnini et al.), we present examples of data on carbon sequestration and biodiversity of tree plantations from field measurements. We also pose the question, how can the farmers who establish and sustain these systems be compensated for their efforts and their contribution to alleviating an environmental problem to society? We describe as an example the program for Payment for Environmental Services in Costa Rica that can serve as a model for other countries with similar ecological and socioeconomic conditions.

Jens B. Aune and collaborators address a similar question in their contribution entitled "Carbon Sequestration in Rural Communities: Is It Worth the Effort?" These authors report data on carbon sequestration potential in agroforestry and forestry projects in Nepal, Uganda and Tanzania. They calculated the economic profitability of these land-use systems, and compared it with the value generated when the carbon was added to the timber value and value of non-wood products. The authors conclude that this increase in value is probably too small to justify the cost in relation to project development, establishment of baseline, carbon monitoring costs, assessment of leakage and documentation of the effect on sustainable development.

The final two articles of this issue, "Shade Coffee Agro-Ecosystems in Mexico: A Synopsis of the Environmental Services and Socio-Economic Considerations," by Sarah Davidson; and "A Review of the Agroforestry Systems of Costa Rica" by Alvaro Redondo Brenes, give an account of traditional and more modern approaches to agroforestry systems, and their environmental services, in two countries of Latin America with long tradition and experience in agroforestry systems, Mexico and Costa Rica. The lessons gained are useful for understanding the complexity of the agroforestry systems and their management, as well as the possibilities for enhancing the environmental benefits provided by the systems.

In order to exploit this vastly unrealized potential of carbon sequestration and protection of biodiversity through agroforestry in both subsistence and commercial enterprises in the tropics and the temperate region, innovative policies such as those described for Costa Rica in this Special Issue have to be put in place. Policies have to be based on rigorous research results, and this Special Issue hopes to contribute with reports of research on environmental services provided by agroforestry systems, and of ways of establishing and managing the systems so that these benefits can be achieved without sacrificing productivity or other goals.

Florencia Montagnini

REFERENCE

Montagnini, F. and P. K. Nair. 2004. Carbon sequestration: An under-exploited environmental benefit of agroforestry systems. *Agroforestry Systems* 61: 281-295.

Soil Carbon Sequestration in Natural and Managed Tropical Forest Ecosystems

R. Lal

SUMMARY. This review article collates and synthesizes the available information on the potential of agroforestry and tropical plantations on soil carbon (C) sequestration to mitigate the greenhouse effect. Tropical forest ecosystems (TFEs) occupy 1.8 billion hectares (Bha) of the total area of 4.2 Bha in forest biomes. The terrestrial C pool in TFEs comprises 120 Mg/ha (tons) in vegetation and 123 Mg/ha in soil to 1-m depth. Soil:vegetation C pool ratio ranges from 0.9 to 1.2 and increases with increase in latitude. Total C pool is 212 petagrams (Pg = 1×10^{15} g = 1 gigaton) in vegetation and 216 Pg in soil. The soil C pool of TFEs represents about 14% of the global soil organic C (SOC) pool of 1550 Pg. Deforestation and conversion of natural to agricultural ecosystems depletes the C pool. Thus, the SOC pool can be enhanced by restoration of degraded soils, and conversion to planted fallows, agroforestry, plantations, improved pastures, and mulch farming. The rate of SOC sequestration in soils is 100-1000 kg C/ha/yr, and total potential of SOC sequestration in TFEs is 200-500 Tg C/yr (1 Teragram = 1^{12} g) for two to five decades. There is a vast potential of converting degraded ecosystems and agriculturally marginal soils to agroforestry and forest

R. Lal is Professor and Director of the Carbon Management and Sequestration Center, The Ohio State University, Columbus, OH 43210 USA.

[Haworth co-indexing entry note]: "Soil Carbon Sequestration in Natural and Managed Tropical Forest Ecosystems." Lal, R. Co-published simultaneously in *Journal of Sustainable Forestry* (Food Products Press, an imprint of The Haworth Press, Inc.) Vol. 21, No. 1, 2005, pp. 1-30; and: *Environmental Services of Agroforestry Systems* (ed: Florencia Montagnini) Food Products Press, an imprint of The Haworth Press, Inc., 2005, pp. 1-30. Single or multiple copies of this article are available for a fee from The Haworth Document Delivery Service [1-800-HAWORTH, 9:00 a.m. - 5:00 p.m. (EST). E-mail address: docdelivery@haworthpress.com].

Available online at http://www.haworthpress.com/web/JSF

doi:10.1300/J091v21n01_01

plantations to restore ecosystems, sequester carbon, and mitigate the greenhouse effect.

KEYWORDS. Soil carbon dynamics, tropical soils, soil degradation, forest ecosystems, soil restoration, greenhouse effect, fossil fuel offset, global warming

INTRODUCTION

The global carbon (C) budget shows that about 3.3 Pg (petagram = 1×10^{15} g = 1 billion ton = 1 gigaton) of C is accumulating in the atmosphere annually (IPCC, 2001). The rate of increase is 1.5 to 2.0 parts per million (ppm) per year, with a high rate of 3 ppm for 2003. Terrestrial C sequestration, being a natural process, is one of the possible strategies for reducing the rate of enrichment of atmospheric CO_2. Large land area and high biodiversity of TFEs warrant a detailed study of their importance in the global C cycle.

The TFEs occur within the humid tropics or the bioclimates characterized by consistently high temperatures and high relative humidity. Total annual rainfall of these regions ranges from 1500 mm to 4500 mm received over 8-12 months. The TFE biome occupies a total area of 1.8 billion hectares (Bha); the vegetation of the humid tropics is dominated by rainforest, covering 1.1 Bha to 1.5 Bha, or about 30% of the land area within the tropics (Table 1). Bruenig (1996) estimated the area of rain-

TABLE 1. Estimates of area under tropical rainforest (adapted from NRC, 1993; FAO, 2003).

	Tropical rainforest area (Mha)		
Region	1980	1990	2000
Africa	289.7	241.8	224.8
Latin America	825.9	753.0	718.8
Asia	334.5	287.5	187.0
Total	1450.1	1282.3	1130.6

forest at 1.64 Bha in 1985 and 1.5 Bha in 1995. Predominant soils of these ecoregions are Oxisols, Ultisols, Alfisols, and Inceptisols (Table 2). Of the total land area of TFEs of 1.8 Bha, 35% are Oxisols, 28% are Ultisols, 15% are Inceptisols, 14% are Entisols, 4% are Alfisols, 2% are Histosols, and 2% comprises Spodosols, Mollisols, Vertisols and Andisols (WRC, 1993). Soil-related constraints to crop production include nutrient imbalance characterized by low availability of N, P, Ca and Mg; low pH, and toxic concentrations of Al and Mn.

The objective of this article is to review the importance of agroforestry, plantations, and other land use and management systems which may restore or enhance SOC pool, improve soil quality and reduce the rate of enrichment of atmospheric concentration of CO_2.

ORGANIC CARBON POOL IN SOILS OF TROPICAL FOREST ECOSYSTEMS

Soil organic carbon (SOC) pool plays an important role in productivity and sustainable use of soils of TFEs through the moderation of cation exchange capacity (CEC), water holding capacity, soil structure, resistance against erosion, nutrient retention and availability and buffering against sudden fluctuations in soil pH. All other factors (clay content, landscape position, drainage, etc.) remaining the same, SOC concentration in soils of the tropics is similar to that of the temper-

TABLE 2. Predominant soils of tropical rainforest ecosystems (adapted from NRC, 1993).

	Area (Mha)			
Soil order	Asia	Africa	Latin America	Total
Oxisols	14	179	332	525
Ultisols	131	69	213	413
Alfisols	15	20	18	53
Inceptisols	90	75	61	226
Entisols	31	91	90	212
Histosols	23	4	–	27
Spodosols	6	3	10	19
Mollisols	7	–	–	7
Vertisols	2	2	1	5
Andisols	1	1	–	2
Total	379	444	666	1489

ate ecosystems (Table 3). The mean SOC pool (kg/m^2) for 0-15 cm and 0-100 cm depths, respectively, has been reported to be 3.8 and 11.3 for Oxisols, 2.9 and 6.4 for Alfisols, 2.1 and 6.4 for Ultisols of the tropics compared with 3.3 and 10.1 for Mollisols, 2.8 and 5.8 for Alfisols and 2.4 and 4.2 for Ultisols of temperate regions (Sanchez et al., 1982). The SOC pool in soils of the lowland tropics is 4-6 kg/m^2, which decreases rapidly to 1-3 kg/m^2 in cropland and 2-4 kg/m^2 under plantation (Woomer et al., 1994). Cerri et al. (2000) reported that the SOC pool in soils of the Amazon is 2.3-21.7 kg C/m^2 to 1-m depth. The mean C pool in TFE is 121 Mg/ha for vegetation and 123 Mg/ha for soil, with a total C pool of 212 Pg in vegetation and 216 Pg in soil worldwide (Dixon et al., 1994). Prentice (2001) estimated the terrestrial C pool in TFE at 553 Pg comprising 340 Pg in vegetation and 213 Pg in soil with corresponding values of 120-194 Mg/ha in vegetation and 120-123 Mg/ha in soil. Thus, the SOC pool of TFEs is about 14% of the global SOC pool of 1550 Pg.

Major differences in the SOC pool in soils of the tropics versus temperate climate lie in the rate of decomposition, land use conversion and soil management. Over and above the differences in chemical composition, the rate of decomposition can be four times faster in the tropics than in temperate climates because of high temperatures (Jenkinson and Ayanaba, 1977). There are also differences in the mechanisms of protection of the SOC pool (Figure 1). Some soils with variable charge are richer in aliphatic (colloidal material of volcanic origin) material and carboxyl groups (Oades et al., 1989). Organic materials are absorbed on oxide surfaces leading to formation of stable micro-aggregates comprising organo-variable-charge-clay systems. It is this microstructure that stabilizes SOC in allophonic soils for several thousands of years (Wada and Aomine, 1973). However, intensive tillage can disrupt these stable aggregates (Wada, 1985; 1986) and release SOC. Strong aggre-

TABLE 3. Mean soil organic carbon content of 61 soils from the tropics and in US soils from temperate regions (modified from Sanchez et al., 1982).

	SOC concentration (%)		
Depth (cm)	Tropical soils	Temperate soils	Significance
0-15	0.97	0.95	NS
0-50	0.64	0.60	NS
0-100	0.40	0.36	NS

FIGURE 1. Mechanisms of protection of organic matter in soils of the tropics.

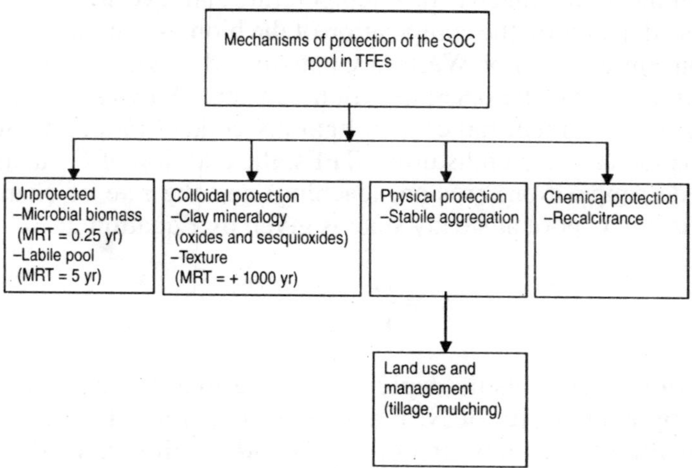

gation is also observed in some soils of eastern Africa and Central and South America. Distribution of SOC deep in the sub-soil, away from the zone of anthropogenic perturbations, and formation of chemically and biologically recalcitrant fractions is also important to retention of C in soil. Charcoal, formed by natural or managed fires, similarly retains C. Thus, the ability of the soils and management systems to protect the SOC pool against anthropogenic perturbations is a key determinant of C sequestration.

CARBON DYNAMICS IN SOILS OF THE TROPICS

The SOC dynamics is described by addition and decomposition of biosolids (Equation 1):

$$\frac{dC}{dt} = -KC + A \qquad (1)$$

where dC/dt is the rate of change of C as SOC pool, t is time, K is decomposition constant and A is accretion of biomass comprising the amount of C added to the soil through crop residue, leaf litter, root biomass and detritus material. Soil C sequestration happens when the quantity (A−KC) is positive and depletion occurs when it is negative.

The factor K is generally higher for tropical than for temperate climates because of the higher mean annual temperature and favorable soil moisture. It also depends on the composition of the biomass, including N and lignin concentrations. For Western Ghats in India, Kumar and Deepu (1992) reported that the decomposition constant (K) was strongly affected by the N content rather than lignin:N content ratio. At equilibrium, as is the case for undisturbed TFEs, the addition of C equals the loss (by decomposition, erosion, leaching) and the rate of change is zero. Thus, the C pool at steady state is given by Equation 2:

$$C = \frac{A}{K} \tag{2}$$

The functional relationship depicted in Equations 1 and 2 is also influenced by availability of N, P and other nutrients. Deforestation and harvesting of biomass in croplands and grazing lands depletes the SOC pool. The rate of SOC depletion upon conversion from natural to agricultural ecosystems is higher in TFEs than in soils of higher latitudes, and it depends on the clay content, degree of aggregation and the frequency and intensity of disturbance. SOC depletion is exacerbated by accelerated soil erosion and other degradative processes.

Conversion of natural TFEs to agricultural land use leads to a rapid decline in the SOC pool which, in severely degraded soils, may decrease to 20% of the antecedent pool (Figure 2). Adoption of recommended management practices (RMPs) on degraded soils can help sequester more SOC. These practices include: no-till cropping of root or grain crops with crop residue mulch and integrated management of soil fertility, adoption of agroforestry measures, establishing plantations (cocoa, coffee) with companion shade crops, and afforestation with rapidly growing and site-adapted plantations (Figure 2). The rate of SOC sequestration in these restorative strategies depends on the amount and quality (C:N ratio, lignin content, etc.) of biomass added, depth and proliferation of the root system, conservation-effectiveness of these measures for erosion control and change in soil moisture and temperature regimes that decreases the rate of decomposition of the biomass. The strategy is to select land use and soil management systems that increase biomass addition to the soil and decrease the rate of its decomposition, so that the quantity (A − KC) in Equation 1 is positive and large.

FIGURE 2. A schematic representation of the dynamics of soil organic carbon in TFEs. The rate of increase in the SOC pool depends on the restorative land use. The rotation time is 12 years for fast-growing species and 20-25 years for slow-growing species. The symbol Δ on each curve denotes the rate (ΔY/ΔX) of SOC sequestration, and it depends on the reference point or base line. The rate may be high and positive when degraded cropland is used as a reference point, and slow or negative when natural TFE is chosen as reference. Afforestation of degraded agricultural soils with rapidly growing plantations may have SOC sequestration rate of 1 Mg C/ha/yr.

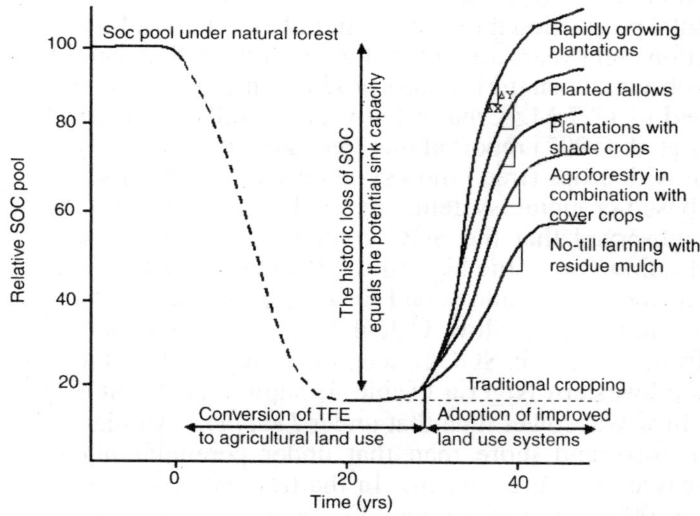

LAND USE AND MANAGEMENT SYSTEMS

The principal land uses in TFEs discussed here are planted fallows, agroforestry, pastures, short rotation tree plantations and root crops based systems. The SOC pool and its dynamics depend on the specific management system.

Planted Fallows

Planted fallows are cover crops specifically established to restore degraded soils and ecosystems, and generally follow the rotation cycle. Fast-growing annuals, herbaceous species and trees can produce a large quantity of biomass which increases the SOC pool and improves soil structure. Ghuman and Lal (1990) assessed the litter fall under planted fallows on an Ultisol in southern Nigeria. The litter fall under *Cassia siamea* was 5.4 Mg/ha and 6.8 Mg/ha two and three years after planting, respectively. The biomass added contributed 113, 91, 13 and 40 kg/ha/yr of N, Ca, Mg and K, respectively. The rate of biomass production for different species grown in southwestern Nigeria ranged from 7-9 Mg/ha/yr three years after planting and 10-14 Mg/ha/yr eight years after planting (Table 4). These rates were comparable to those of biomass production in the bush fallow system. The litter fall comprised 74-96% of the total production three years after planting and 82-91% of the total production eight years after planting. In southern Senegal, Manlay et al. (2002) observed that the biomass production of 5.5 Mg C/ha in cropland increased to 17.7 Mg C/ha in ten years of fallowed plots. In Zambia, Barrios et al. (1997) reported that trees grown for two to three years in rotation with crops (tree fallows) increased soil fertility in maize *(Zea mays)* based cropping systems. In southwestern Nigeria, Salako et al. (1999) observed that the SOC contents of the 0-5 cm layer under planted fallows were the highest for *Pueraria phaseoloides* for 1-year cropping and 2-year fallow, and for *Leucaena leucocephala* for 1-year cropping and 3-year fallow. Calculations of the SOC pool for 0-15 cm depth from the data on SOC concentration reported by Juo et al. (1995) from southwestern Nigeria (Table 5) show that the SOC pool under bush fallow was lower than that under *Leucaena*, similar to that under Guinea grass and more than that under perennial pigeon pea and maize-based cropping systems. In the tropical Andes, Sarmiento and Bottner (2002) reported that the SOC pool was maintained under a fal-

TABLE 4. Biomass production by a range of planted fallows grown on a degraded Alfisol in western Nigeria (adapted from Salako and Tian, 2001).

Species	Total biomass production	
	3 year growth rate	8 year growth rate
	kg/ha/yr	
Senna siamea	7,779 (91.5)	10,368 (82.1)
Leucaena leucocephala	8,783 (74.0)	10,049 (83.4)
Acacia leptocarpa	7,602 (83.3)	10,665 (90.7)
Acacia auriculiformis	6,919 (96.1)	12,107 (86.9)
Bush fallow	7,714 (87.6)	13,612 (84.0)
Hedgerows	8,250	11,320

Numbers in parenthesis refer to litter fall as % of the total biomass.

TABLE 5. Temporal changes in SOC stock of 0-15 cm depth of an Alfisol in western Nigeria with cultivation and fallowing treatments (recalculated from Juo et al., 1995).

Years after clearing	Bush fallow	Guinea grass	Leucaena	Pigeon pea	Maize + stover	Maize− stover
	--------------------	--------------------	Mg C/ha	--------------------	--------------------	--------------------
0	22.5	28.2	22.9	23.9	29.0	24.1
4	18.0	20.8	19.1	20.0	18.5	14.4
7	15.1	17.1	18.5	16.0	18.2	12.8
10	27.4	31.2	35.1	25.2	27.1	24.8
12	27.4	30.0	33.0	20.8	28.1	27.0
13	30.0	30.9	34.3	25.2	29.8	26.0

low system and did not decrease further. In Brazil, Denich et al. (2000) observed that fallow management is a good strategy for SOC sequestration by small-scale farmers who cannot afford chemical fertilizers.

It is apparent, therefore, that planted fallows are important to maintaining the SOC pool at an optimum level for sustainable use of soils of the TFEs. Determining the frequency of incorporating planted fallows in the rotation cycle is important to the sustainability of the cropping system. The duration of cropping (t_c) and planted fallows (t_{pf}) can be adjusted to attain the desired level of SOC. If the mean SOC content of the soil is C_m, Equation 1 can be re-written as follows:

$$(-K_c C_m + A_c)t_c + (-K_{pf} C_m + A_{pf})t_{pf} = 0 \qquad (3)$$

where K_c and K_{pf} refer to decomposition constants, A_c and A_{pf} are accretion constants, and t_c and t_{pf} are the duration of cultivation and planted fallows, respectively. Rearrangement of Equation 3 allows computation of the ratio of cropping periods to the planted fallow period to maintain the desired level of SOC.

$$\frac{t_c}{t_{pf}} = \frac{(A_{pf} - K_{pf} C_m)}{(K_c C_m - A_c)} \qquad (4)$$

where K and A are determined experimentally and the desired level of C_m can be fixed for a specific ecoregion.

Agroforestry Systems

It is widely recognized that establishing trees in degraded ecosystems improves soil quality (Buresh and Tian, 1998). Therefore, growing trees in association with crops or pastures can also improve soil quality or reduce the risks of soil degradation. Agroforestry is a strategy of growing trees/woody perennials and crops or pastures on the same land at the same time. There are numerous types of agroforestry systems (Sanchez, 1995; Sanchez et al., 1994), and these are useful because some can duplicate important characteristics of undisturbed ecosystems (Tornquist et al., 1999). Agroforestry systems which increase the biomass returned to the soil and decrease the rate of its decomposition make the quantity (A – KC) in Equation 1 positive and enhance the SOC pool. The SOC pool of the cropland increases only if either the addition of the biomass is enhanced (A) and/or the decomposition rate (K) is decreased (Sauerbeck, 2001). Site-specific agroforestry systems provide opportunities by which such improvements can be achieved. Wright et al. (2001) compared C sequestration potential of a maize-based system with that of an agroforestry system to sequester C at global scale. They estimated that an additional 670-760 Mha of maize area under improved management would be required to assimilate 3.3 Pg C yr^{-1} being added annually into the atmosphere (Wright et al., 2001). In contrast, adoption of agroforestry on just 460 Mha would achieve the same objectives (Wright et al., 2001). Wright and colleagues argued that agroforestry is the only system that could realistically be implemented to reduce global CO_2 levels through terrestrial C sequestration, a claim that needs to be verified and critically assessed.

Alley cropping, or hedgerow cropping, is a system of agroforestry where trees and crops are intercropped, the former being periodically pruned to produce mulch and minimize the adverse impact of shading (Haggar et al., 1993). Beneficial effects of alley cropping on improvements in soil fertility and crop yield have been reported from Ghana (Astivor et al., 2001), and the sub-humid highlands of Kenya (Mugendi et al., 1999). In Ghana, Astivor and colleagues reported that mean SOC concentration at 0-15 cm depth was 18.6 g/kg under eight-year old woodlot of *Leuçaena leucocephala*, 14.0 g/kg under alley cropping, 14.5 g/kg under natural fallow and 13 g/kg under conventional tillage. In Kenya, Mugendi et al. (1999) reported that *Leucaena* hedges contributed more prunings and N than those of *Calliandra*. However, these hedges decreased yield of maize due to competition for light, nutrients and water. For an Ultisol in southern Nigeria, Ghuman and Lal (1990) reported that the biomass production by hedgerows of *Gliricidia sepium* in an alley cropping system contributed 10.8 Mg/ha/yr of biomass containing 2.3 Mg/ha/yr of leaves and 81.2 kg/ha/yr of N. Similar observations have been reported by Kang (1987) and Mulongoy and van der Meersch (1988). Kang (1997) reported that SOC concentration at 0-15 cm depth after establishing five years of *Leucaena* hedgerows was 12.3 g/kg under the hedgerows, 9.4 g/kg in the alleys between the hedgerows, and 5.9 g/kg in the control without hedgerows. The data in Table 6 shows the SOC pool under different cropping systems on an Alfisol in southern Nigeria. Over a five year period, the SOC pool declined in all ecosystems. The mean rate of decline in the 0-10 cm layer over the five-year period was 3.0 Mg C/ha/yr in plow till, 1.2 Mg C/ha/yr in no-till, 1.4 Mg C/ha/yr in *Leucaena* hedgerows 4 m apart, 1.9 Mg C/ha/yr in *Leucaena* hedgerows 2 m apart, 2.2 Mg C/ha/yr in *Gliricidia* hedgerows 4 m apart, and 2.4 Mg C/ha/yr in *Gliricidia* hedgerows 2 m apart.

In northern India, Chander et al. (1998) assessed the impact of intercropping of wheat (Triticum aestivum) and cowpeas (Vigna unguiculata) with 12-year old N_2-fixing *Dalbergia sissoo*. They observed that the SOC and microbial biomass C increased in soils under *D. sissoo*. Sev-

TABLE 6. Alley cropping effects on soil organic carbon pool of 0-10 cm depth under maize-cowpea rotation for an Alfisol in western Nigeria (recalculated from Lal, 1989a; b; c).

	Years after establishment of tree hedgerows				
Treatment	1	2	3	4	5
	Mg C/ha				
Plow	20.0	14.6	10.8	10.6	5.2
No-till	18.9	20.4	11.2	16.2	13.1
Leucaena-4 m	20.5	22.6	17.1	19.9	13.6
Leucaena-2 m	20.6	17.6	19.3	15.5	11.1
Gliricidia-4 m	22.0	22.2	14.6	16.4	11.0
Gliricidia-2 m	20.4	13.8	11.7	17.2	8.2

eral agro-forestry systems used in southeastern Asia have reportedly improved the SOC pool. In the Philippines, Malab et al. (1996; 1997) reported that hedgerows of Desmanthus virgatus increased SOC concentration in the surface layer from 11.4 g/kg to 15.3 g/kg after four years of continuous mulching produced by the prunings. Lasco (1991) evaluated biomass production and soil properties for three species grown as hedgerows at Los Baños, Philippines: *Leucaena leucocephala*, *Sesbania grandiflora* and *Pennisetum purpureum*. Mulching with the herbage significantly improved the SOC pool, total soil N, and available P and K. In Thailand, Patma-Vityakon (1991) reported that the SOC concentration was high in systems with frequent return of biomass to the soil.

Several studies conducted in Africa indicate the importance of agroforestry systems on erosion control (Lal, 1989a). Roose and Barthes (2001) reported that the rate of SOC loss by erosion and leaching ranged between 10 and 100 kg C/ha/yr on sloping cultivated lands in west Africa. Established perennial hedgerows in Rwanda reduced losses from runoff and erosion, and observed that agroforestry systems which produce good-quality litter are an important part of the solution. *Leucaena* hedgerow intercropping systems with cattle manure have also been successful in Ethiopia (Lupwayi and Haque, 1999a; b; Lupwayi et al., 1999). In Kenya, Maroko et al. (1998) observed tree-root competition with maize can be minimized through species selection. Barrios et al. (1996a; b; 1997) reported improvement in soil fertility and N-availability by using prunings of *Sesbania* and *Gliricidia*. Woomer et al. (2000) reported that a 23-year old agroforestry system had a total system C pool (including plant components) of 130 Mg C/ha compared to 48 Mg C/ha in a pastoral system.

Agroforestry systems have been widely used in South and Central America. In Brazilian Amazonia, Schroth et al. (2002) observed that 7 years after establishment, multi-strata systems had an above-ground biomass of 13.2-42.3 Mg/ha, a below-ground biomass of 4.3-12.9 Mg/ha and a litter mass of 2.3-7.2 Mg/ha compared with those for monoculture at 7.7-56.7 Mg/ha, 3.2-17.1 Mg/ha and 1.9-5.6 Mg/ha, respectively. The combined biomass and litter was the highest in peach palm (*Bactris gasipaes*) for fruit. In comparison, 14-year old secondary forest had a combined biomass and litter stock of 127 Mg/ha. At Yurimaguas, Peru, Woomer (1993) reported that the SOC pool in the surface layer was 2.7 kg/m^2 for rainforest, 2.8 kg/m^2 for field crop and 2.6 kg/m^2 for an agroforestry system. In the Atlantic region of Costa Rica, Tornquist et al. (1999) studied the impact of tropical hardwoods (*Vochysia ferruginea*, *Vochysia guatemalensis*, *Stryphnodendron microstachyum* and *Hieronyma alchorneoides*) as a tree component of an agroforestry system and compared soil properties with pasture systems. Higher mineralizable C levels were observed in pastures than under agroforestry. Also in Costa Rica, Mazzarino et al. (1993) observed no significant differences in the SOC pool of a Eutric Cambisol after nine years of alley cropping and control, although the SOC concentration increased by 10% under alley cropping over this period. Kass et al. (1989; 1997) also studied the impact of agroforestry systems using *Erythrina poeppigiana* and *Gliricidia sepium*. They observed that for systems with low nutrient removal trees that are allowed to grow for long periods of time without being pruned would make greater contributions to the improvement of soil quality than those in which frequent pruning is the normal practice. In the French Antilles, Dulormme et al. (1999) reported that the SOC concentration after ten years of agroforestry increased from 2.41% to 2.84% at 0-10 cm depth, and from 1.83% to 2.16% at 10-20 cm depth.

The sign and magnitude of the term (A−KC) for an agroforestry system determines the dynamics of the SOC pool and management-induced changes in soil properties. The magnitude of (A−KC) depends on species of tree/shrub, soil type, rainfall, climate and the management. Differences in the site-specific factors lead to differences in soil and crop response to agroforestry systems. The term (A−KC) in Equation 1 was negative in all systems in Table 6 because the biomass returned was not enough to balance the decomposition rate. Incorporating *Mucuna utilis* in the rotation cycle along with a combination of no-till and a hedgerow system may have made the term (A−KC) positive and maintained or enhanced the SOC pool. For example, in Pakistan eucalyptus + wheat-based agroforestry systems decreased the SOC pool in several soils (Bhatti and Khan, 2002), but improved crop yield and soil quality in Sudan (El-Amin et al., 2001). Such contradictions may need to be resolved by using ^{13}C analyses and modeling techniques (Diels et al., 2001), in conjunction with an evaluation of the effects of soil type, nutrient availability, soil moisture regime, species, and the biomass returned.

Pastures

Pastoral land use is an economically important agroecosystem in the TFEs. Yet conversion of TFE into pasture leads to release of CO_2 and other greenhouse gases into the atmosphere. The undisturbed TFE may contain as much as 406 Mg of biomass/ha of which 309 Mg may be the above-ground material (Fearnside, 2000). Thus, deforestation and burning of TFE leads to large emissions of CO_2 on a global scale (IPCC, 2001). The establishment of pastures can improve (Cerri et al., 1991; 1994; Moraes et al., 1995; 1996; Neill et al., 1997; Noordwijk et al., 1997), deplete (Trumbore et al., 1995; Desjardin et al., 1994; Luizao et al., 1992) or have no effect (Feigl, 1995) on the SOC pool, depending on the soil type and antecedent SOC pool, pasture species and management. Neill and Davidson (2000) reported that nineteen of the 29 pastures examined in the Amazon accumulated C in surface soils and ten showed C loss. Pastures formed on forest soils with high antecedent SOC pool tended to lose C, while those formed on soils with low SOC pool (< 5 kg/m^2) tended to gain SOC. Pastures planted with *Brachiaria humidicola* tended to lose SOC; those planted with *Panicum maximum* and *Brachiaria brizantha* tended to gain SOC. In the Cerrado region of Brazil, Resck et al. (2000) reported that decomposition rates are high especially in loamy Latosols and quartz sand, and that managed pastures have a high capacity to recover the degraded SOC pool and improve soil physical quality. In the Atlantic region of Costa Rica, Veldkamp (1994) observed a very slow decrease in the SOC pool under pasture. In the acid soil savanna region of Colombia, Fisher et al. (1994) reported drastic increase in the SOC pool by growing deep-rooted grasses. Some planted fallows can be used as pastures, and SOC under pastures can be greatly enhanced with controlled grazing, establishing improved species and fertility management (Koutika et al., 1999; 2000). See Amezquita et al. (this volume) for more on this subject.

Plantations

In addition to C sequestration in biomass and soil, tropical plantations are needed for timber, and more importantly, as fuel wood for cooking. Thus, the area under tropical plantations has increased drastically since the 1960s, from 7 Mha in 1965 to 21 Mha in 1980, 43 Mha in 1990 (Evans, 1992) and 187 Mha in 2000 (FAO, 2003). Despite the rapid expansion, tropical plantations occupy a relatively small area in relation to other land uses. Extensive literature supports the conclusion that afforestation of degraded soils increases the SOC pool with accompanying improvement in soil quality. Tropical plantations cause less soil disturbance, generally involve no prunings and return a large quantity of leaf litter and detritus material to the soil. Consequently, the SOC pool is either maintained or enhanced under plantations.

Tree plantations are intensively managed tracts of land that are populated by fast-growing single or mixed species to maximize timber or fuelwood production. Establishment of such plantations on degraded lands has large potential for terrestrial C sequestration. It is important, however, to use site-adapted and well-domesticated trees (Bruenig, 1996; Leaky and Newton, 1994). Commonly used species of relatively fast-growing trees in TFEs include: *Acacia mangium, Albizia* spp., *Anthocephalus chinensis*, *Carapa guianensis*, *Casuarina* spp., *Dalbergia nigra*, *Dinizia excelsa*, *Eucalyptus* spp., *Euxylophora paraensis*, *Gmelina arborea*, *Leucaena* spp., *Octomeles sumatrana*, *Paraserianthes falcataria*, *Parkia multijuga*, *Pinus caribaea*, *Prosopis* spp., *Pterocarpus* spp., *Shorea javanica*, *Swietenia machophylla*, *Tectona grandis*, among others. Also, there are large areas in the tropics of plantations of tree crops such as oil palm, rubber, coffee, cashew and cocoa. Establishment of such plantations is a useful strategy whenever natural succession is not effective in rehabilitating degraded ecosystems (Appanah and Weinland, 1992). Timber trees and tree crops sequester C in soil and biomass when established on degraded soils and ecosystems, and nutrient losses during site preparation can be minimized through reducing soil disturbance, minimizing erosion and avoiding burning (Nykvist et al., 1994).

In western Nigeria, Ekanade et al. (1991) reported that the SOC pool under forest was 29 g/kg and that under cocoa was 19 g/kg. Similar observations were made by Adejuwon and Ekanade (1988) in Oyo state, Nigeria. Also in southern Nigeria, Ojunkunle and Eghaghara (1992) observed that the SOC concentration under 10-year old cocoa plantation was 25 g/kg compared with 35 g/kg under forest. In Nigeria, Aweto (1987) reported that the SOC concentration was 14 g/kg under primary forest and 12 g/kg under a 18-year old rubber plantation. The SOC concentration under rubber increased over time. In Kade, Ghana, Duah-Yentumi et al. (1998) reported that the SOC concentration of a soil under 40-year old rubber plantation was lower than that under virgin forest or 20-year old cocoa. Both rubber and cocoa received no fertilizer or manure.

At Turriabla, Costa Rica, SOC concentration under cocoa increased from 28 g/kg to 32 g/kg in 0-15 cm and 23 g/kg to 25 g/kg in 15-30 cm depth in 9 years when cocoa was shaded with *Erythrina poeppigiana* (Beer et al., 1998). Similar improvements in SOC concentration were observed when the cocoa was shaded with *Cordia alliodora* (Beer et al., 1998).

On the basis of a 16-year study on a rubber plantation in Malaysia, Sanchez et al. (1985) reported that the SOC concentration decreased from 18 g/kg under forest to 10 g/kg under rubber. The SOC concentration increased when *Pueraria* was grown under rubber. In Malaysia, Pushparajah (1998) reported that the SOC concentration increased after 20 years of oil palm plantation. In Sumatra, Indonesia, Lumbanraja et al. (1998) reported that the SOC concentration in degraded cropland and primary forest, respectively, was 15.0 g/kg and 60.4 g/kg for 0-20 cm depth, and 7.5 g/kg and 25.0 g/kg for 20-40 cm depth. Conversion of forest to annual crops leads to drastic reductions in the SOC pool in most soils of the TFEs (Hartemink, 2003). Thus, conversion of degraded cropland to plantation would increase the SOC pool over time. In another study, Lumbanrajah et al. (1998) reported that the SOC concentration under 20-year old coffee plantations were 29 g/kg compared with 60 g/kg under forest, and the SOC concentration in cropland were half of that under coffee (15 g/kg) (Table 7). In Papua New Guinea, Kunu and Hartemink (1997) observed no differences in soil chemical properties under coffee and primary forest. Härdter et al. (1997) reported a slight increase in SOC concentration under five-year old oil palm plantation in Malaysia, and indicated that plantations could be established on 11 Mha of degraded land in Indonesia and 1 Mha in Malaysia. Growing legumes in between young plantation trees during the first 5 years is an important strategy for erosion control, nutrient cycling, and SOC sequestration (Beer et al., 1998).

TABLE 7. Changes in SOC concentration by land use conversion in south Sumatra between 1970 and 1990 (adapted from Lumbanraja et al., 1998).

	SOC concentration (g/kg)					
	Bukit Ringgis		Sekinaku		Trimulyo	
Land use	0-20 cm	20-40 cm	0-20 cm	20-40 cm	0-20 cm	20-40 cm
Primary forest	60.4	25.0	73.1	32.7	58.0	18.2
Secondary forest	41.4	21.7	42.9	24.5	38.7	12.5
Coffee plantation	28.5	10.1	30.9	11.2	26.7	7.0
Cropland	15.0	7.5	19.3	7.0	14.2	5.5

Considerable research on C balance under plantations has been done in Costa Rica (Montagnini and Sancho, 1990; Stanley and Montagnini, 1999; Montagnini, 2000). Montagnini and Porras (1998) analyzed soil properties 4 years after establishing plantations. Although there is no control (cropland) for comparison, the data in Table 8 show differences in the SOC pool among species. The maximum SOC pool was measured under *Jacaranda copaia* in Site 1, regeneration in Site 2, and *Balizia elegans* in Site 3. The difference in the the SOC pool to 60-cm depth between the minimum and the maximum over a 4-year period was 16.7 Mg/ha for Site 1, and 24.8 Mg/ha for Sites 2 and 3. Smith et al. (2002) studied the impact of plantation forestry on the SOC pool at the Curua-Una Forest Reserve, Pará, Brazil. Several plantations were established between 1959 and 1973, and soil was sampled in 2000. The data in Table 9 show that the SOC pool in the top 20-cm depth ranged from 72 to 115 Mg C/ha. In comparison with the undisturbed forest, the SOC pool decreased by 27% in *Pinus caribaea* and increased by 16% in *Euxylophora paraensis*. There were also differences in forest floor biomass and fine roots among species. The data on SOC sequestration under plantation at Toa Baja in Puerto Rico are shown in Table 10. The mean rate of SOC sequestration in the 0-20 cm layer under *Albizia*

TABLE 8. SOC pool (Mg C/ha) to 60-cm depth 4 years after planting (calculated from Montagnini and Porras, 1998).

Site 1		Site 2		Site 3	
Species	SOC pool	Species	SOC pool	Species	SOC pool
Jacaranda copaia	90.4	*Albizia guachapele*	75.3	*Genipa americana*	61.1
Calophyllum brasiliense	73.7	*Dipteryx panamensis*	73.1	*Hieronyma alchorneoides*	64.4
Stryphnodendron microstachyum	74.8	*Terminalia amazonia*	70.6	*Balizia elegans*	85.9
Vochysia guatemalensis	76.1	*Virola koschnyi*	71.9	*Vochysia ferruginea*	66.9
Mixed	74.9	Mixed	73.2	Mixed	68.5
Regeneration	85.1	Regeneration	95.4	Regeneration	76.8

Assumptions: Organic matter comprised 58% C, and soil bulk density equals 0.6 Mg/m^3 (Fisher, 1995) for all depths and under all species.

TABLE 9. Effects of plantations of 27-41 years duration on organic carbon pool at 0-20 cm depth for a soil in Curua-Una Forest Reserve, Para, Brazil (adapted from Smith et al., 2002).

Treatment	Soil bulk density (Mg/m^3)	Soil organic carbon: Concentration (g/kg)	Soil organic carbon: Pool (Mg/ha)
Forest	0.77a	63.9ab	98.4ab
Pinus caribaea	0.84a	42.8c	71.9c
Carapa guianensis	0.80a	49.4bc	79.0bc
Leguminosae	0.81a	51.4bc	83.3bc
Euxylophora paraensis	0.82a	69.9a	114.6a

Leguminosae comprised a combination of *Parkia multijuga*, *Dinizia excelsa* and *Dalbergia nigra*.

TABLE 10. SOC and N pools in the 0-20 cm soil layer of Typic Troposamments in control and 4.5 year old plantations of *Albizia lebbek* in Puerto Rico (calculated from Parrotta, 1992).

	SOC		Nitrogen		Sequestration ratio (kg/ha/yr)	
Treatment	Concentration (%)	Pool (Mg C/ha)	Concentration (%)	Pool (Mg N/ha)	SOC	N
Plantation	1.70 (1.04)	35.4	0.095 (1.04)	1.98	1022	89
Control (grasses)	1.44 (1.07)	30.8	0.074 (1.07)	1.58	–	–

Number in parenthesis is soil bulk density in Mg/m^3

lebbek was 1022 kg/ha/yr, which is a very high rate. Several experiments in South America have shown variable response of SOC to replacement of native forest by plantations. In the Cerrado region of Brazil, Zinn et al. (2002) observed that conversion of native forest to plantations decreased the SOC pool at 0-60 cm depth by 11 Mg/ha in *Pinus* over twenty years and 9 Mg/ha under *Eucalyptus* in seven years. The magnitude of SOC depletion was more in soils with low than high clay content. Significant SOC losses under plantation were also reported by Madeira et al. (1989) and Fonseca et al. (1983). In contrast, no effect on the SOC pool under plantation was reported by Lepsch (1980) and Prosser et al. (1993), and increase in the SOC pool was observed by Sanginga and Swift (1992) and Althoff et al. (1996). Such contradictory results for *Pinus* have been reported by Bernhard-Reversat (1996) and Turner and Lambert (2000).

There is a wide range of factors that affect the response of SOC to conversion of native forest to plantation. Important among these are: soil type (clay content and mineralogy), drainage conditions, antecedent SOC pool, species, management and especially the availability of nitrogen and other nutrients. The effect of establishing tree plantations on soil C storage differs among tree species, which differ in biomass production, tissue nutrient concentrations and their effects on soil quality. There is generally 20-100% more SOC under N-fixers than non-N fixers or 0.05-0.12 kg/m^2/yr more C under N-fixers. Resh et al. (2002) attributed this difference to greater retention of older soil C under N-fixing trees. Some plantation species may also grow better under the enhanced CO_2-fertilization effect, assumed to be 0.5-2 Pg C/yr or the equivalent to 8-33% of the annual global fossil fuel emissions (Davidson and Hirsch, 2001).

The effect of tree plantations on SOC sequestration also depends on the management. Poorly managed plantations may as well be a source rather than a sink for CO_2. In Nigeria, Aweto (1995) observed a decline in SOC pool under plantation compared with natural forest in southern Nigeria. The rate of decline in the SOC pool in kg C/ha/yr for 0-20 cm layer was 392 in teak, 492 in gmelina, 627 in cashew, 720 in rubber, 1890 in oil palm and 1144 in coffee. Aweto concluded that "plantations appear to have the potential of contributing towards global warming–a threat they are supposed to mitigate." Once again, the term $(A-KC)$ was negative under plantations, as the vegetation cover was less dense, plants were less diverse, and the biomass C returned to the soil was less than would have occurred under natural forest. For the $(A-KC)$ term to be positive under plantation, the rate of biomass production and return must be more than the natural forest, as was reported for *Pinus caribaea* in Puerto Rico (Cuevas et al., 1991). If plantations in southern Nigeria were established on degraded cropland they could become a sink for CO_2.

Therefore, identification of degraded soils and choice of site-adapted species for a short-rotation plantation is an important strategy for C sequestration. The data in Table 11 show unused/fallow land and some degraded cropland and pastureland in the Brazilian Amazon which can be converted to tree plantations. There also exists a potential for tree plantations in the unused and degraded lands in the Cerrados (Table 12). Similarly degraded soils and ecosystems exist in Latin America, sub-Saharan Africa, and tropical and sub-tropical regions of Asia. Such

TABLE 11. Temporal trends in deforested area and its use in Brazilian Amazonia (adapted from Margulis, 2004).

	Land Use Change (%)				
	1970	1975	1980	1985	1995
Total area deforested (UNITS)	3.0	4.0	6.2	7.7	9.5
Cropland	0.3	0.6	1.0	1.2	1.1
Planted pastures	0.7	1.4	2.6	3.8	6.6
Unused and fallow	2.0	2.0	2.6	2.7	1.8

Original forested area = 419 million hectares (Mha).
Deforested area: 1978 (15.2 Mha); 1990 (41.5 Mha); 2000 (60.3 Mha).

TABLE 12. Trends in land use in the Brazilian Cerrados (adapted from Margulis, 2004).

Land use (Mha)	1975	1996	Average annual growth (%)
Agricultural area	110.8	124.3	0.5
Anthropic area	34.7	64.5	3.0
Fallow area	0.36	0.67	7.4
Productive area not in use	10.8	4.6	−3.9

Anthropic area = crops, planted pastures, reforested, fallow, and productive land not in use.

lands need to be identified as potential sites for terrestrial C sequestration via the judicious management of fast-growing plantations.

Root Crop Based Systems

Cassava (*Manihot esculenta*), yam (*Dioscorea* spp.), sweet potato (*Ipomea batata*) and taro (*Xanthosomas esculenta*) are examples of important root crops for the TFEs. Severe erosion under cassava can deplete the SOC pool. In Colombia, Ruppenthal et al. (1997) reported SOC loss by erosion under cassava at 15-110 kg C/ha/yr with grass barriers, 38-670 kg C/ha/yr with intercropping, 64-330 kg C/ha/yr with flat planting and 4346-7257 kg C/ha/yr with unplanted bare fallow. Root crops require effective 'root room' for tubers to grow; they also respond positively to mulching because of an effective erosion control, water conservation and temperature moderation (Lal, 1975; Hahn et al., 1979). Including restorative cover crops in the rotation cycle and growing root crops as an agroforestry system are important strategies for enhancing the SOC pool, improving soil quality and increasing productivity. In addition, the application of fertilizers and integrated nutrient management can enhance the SOC pool (Jin et al., 1999) in root crop systems.

GLOBAL POTENTIAL OF SOC SEQUESTRATION IN TFES

There are three strategies of SOC sequestration in TFEs: (i) establishing of tree plantations on degraded and agriculturally marginal soils, (ii) incorporating planted fallows in the rotation cycle, and (iii) using agroforestry techniques. Afforestation, the establishment of tree plantations after ≥ 50 years without forest, has a large potential for SOC sequestration in the tropics (Table 13). The area that must be afforested with tropical plantations to sequester 1 Pg C/yr by 2054 is estimated to be 250 Mha (Pacala and Socolow, 2004). High rates of SOC sequestration at 0.8-4.0 Mg C/ha/yr have been reported for secondary forest succession on land that had been cultivated for 100-300 years in Puerto Rico (Lugo and Sanchez, 1986). Bouwman and Leemans (1995) estimated that afforestation stored 50 Mg of SOC in 30 years. Johnson (1992) reported a > 35% increase in soil C following the afforestation and reforestation of cultivated soils. Brown and Lugo (1990) reported SOC increase of 1-2 Mg C/ha/yr in the wet tropical zone. The data by Parrotta et al. (1992) show SOC sequestration rate of 1022 kg C/ha/yr (Table 10). The rates of SOC sequestration are generally lower in the dry tropics than in the wet tropics. Deans et al. (1999) reported SOC accumulation under 18-year plantation of *Acacia senegal* in northern Senegal at the rate of 0.03%/yr under the tree canopy and 0.02%/yr in the open ground, corresponding to SOC sequestration rates of 420 and 280 kg C/ha/yr for a soil bulk density of 1.4 Mg/m^3. In contrast, Bashkin and Binkley (1998) reported no or little net increase in net SOC sequestration following afforestation.

The CO_2 fertilization effect also has the potential to increase SOC sequestration. However, the positive response to CO_2 fertilization may be limited by the availability of N and other essential nutrients (Schlesinger and Lichter, 2001; Oren et al., 2001). The data in Table 14 show an estimate of the potential of SOC sequestration in TFEs. Thus, the total potential of SOC sequestration in these ecosystems is 200 to

TABLE 13. Land area in tropical regions potentially available for afforestation and adoption of recommended management practices for soil carbon sequestration (adapted from Grainger, 1988; Schroeder, 1992).

Land use	Africa	Asia	Latin America	Total
	-----------	-----------	Mha -----------	-----------
Logged forests	39.0	53.6	44.0	136.6
Forest fallows	59.3	58.8	84.8	202.8
Deforested watersheds	3.1	56.5	27.2	86.9
Desertified drylands	740.9	748.0	162.0	1650.9
Total	842.3	916.9	318.0	2077.2

TABLE 14. The potential of soil organic carbon sequestration in the TFE of the humid tropics.

Land use	Area (Mha)	Rate of SOC sequestration (kg C/ha/yr)	Potential of SOC sequestration (Tg C/yr)
Agroforestry	500	100-300	50-150
Plantations	250	500-1,000	125-250
No-till mulch farming	50	100-200	5-10
Improved pastures	200	100-500	20-100
Total	1,000		200-510

500 Tg C/yr. The sink capacity may be filled over two to five decades, provided that the restorative land use is followed on a continuous basis. Whereas the rates of SOC sequestration used in these computations are supported by the literature reviewed, the projections of land areas that can be converted to improved and restorative management strategies are based on the estimates by Grainger (1988) and Schroeder (1992). Granger (1988) estimated that tropics contain over 2 Bha of degraded and depleted lands. Of these, 758 Mha were once forested and include 137 Mha of logged tropical moist forests, 203 Mha of forest fallows, 87 Mha of deforested watershed areas and 331 Mha of rainfed/irrigated cropland (Granger, 1988). These areas are suitable for afforestation and adoption of recommended management practices.

CONCLUSIONS

Sustainable management of tropical biomass is essential to terrestrial C sequestration and improving soil and environment quality. Avoiding deforestation through prudent management for land already cleared is important. Appropriate land uses for enhancing the terrestrial C pool include: planted fallows, plantations, site-adapted pastures with deep root systems, and agroforestry systems. The rate of SOC sequestration in TFEs may be 100 to 1000 kg C/ha/yr. High rates may be obtained on clayey soils with favorable soil moisture regimes in farming systems that involve minimal soil disturbance, ensure a continuous supply of biomass to the soil surface, and maintain favorable nutrient balance within the soil and ecosystem. Establishing species adapted to the soil-specific conditions, which produce a large quantity of biomass and have a deep and prolific root system, supports high rates of SOC sequestration. Formation of stable micro-aggregates or organo-mineral complexes is another important mechanism of protection of SOC against microbial processes. The SOC encapsulated within aggregates is not easily decomposed. Site-specific choice of an appropriate farming system depends on soil properties, climate, terrain, and socio-economic, ethnic and cultural factors. Tree-based systems, with planted fallows as shade and mulching, are important components that will contribute to achieving the sustainable use of soil and water resources in TFEs. Soil fertility enhancement through nutrient management is also important to SOC sequestration. Realizing the total potential of SOC sequestration (200-500 Tg C/yr) will necessitate implementation of appropriate policies that encourage adoption of restorative land uses and quantification of changes in the SOC pool over time, so that C credits can be traded in domestic and international markets.

REFERENCES

Adejuwon, J.O. and O. Ekanada. 1988. A comparison of soil properties under different land use types in a part of the Nigerian cocoa belt. *Catena* 15: 319-331.

Althoff, P.P., K. MacDicken and D. Chandler. 1996. Comparative inventory of sequestered carbon in a plantation of *Eucalyptus camaldulensis* and in 17-year-old natural regeneration in Brazil's *Cerrado*. In: Proceedings of the International Symposium on Forest Ecosystems, Vol. VI. Belo Horizone, Brazil, pp. 170-171.

Appanah, S. and G. Weinland. 1992. Will the management systems for hill Dipterocarp forests stand up? *J. Trop. For. Sci.* 3: 140-158.

Astivor, L., G.N. Dowuona and S.G.K. Adiku. 2001. Farming system-induced variability of some soil properties in a sub-humid zone of Ghana. *Plant Soil* 236: 83-90.

Aweto, A.O. 1987. Physical and nutrient status of soils under rubber (*Hevea brasiliensis*) of different ages in southwest Nigeria. *Agric. Systems* 23: 63-72.

Aweto, A.O. 1995. Organic carbon diminution and estimates of CO_2 release from plantation soil. *The Environmentalist* 15: 10-15.

Barrios, E., R.J. Buresh and J.I. Sprent. 1996a. Nitrogen mineralization in density fractions of soil organic matter from maize and legume cropping systems. *Soil Biol. Biochem.* 28: 1459-1465.

Barrios, E., R.J. Buresh and J.I. Sprent. 1996b. Organic matter in soil particle size and density fractions from maize and legume cropping systems. *Soil Biol. Biochem.* 28: 185-193.

Barrios, E., F. Kwesiga, R.J. Buresh and J.I. Sprent. 1997. Light fraction soil organic matter and available nitrogen following trees and maize. *Soil Sci. Soc. Am. J.* 61: 826-831.

Bashkin, M.A. and D. Binkley. 1998. Changes in soil carbon following afforestation in Hawaii. *Ecology* 79: 828-833.

Beer, J., A. Bonneman, W. Chavez, H.W. Fassbender, A.C. Imbach and I. Martel. 1990. Modelling agroforestry systems of cocoa (*Theobroma cacao*) with laurel (*Cordia alliodora*) or poro (*Erythrima poeppigiana*) in Costa Rica. V. Productivity indices, organic material models and sustainability over 10 years. *Agroforestry Systems* 12: 229-249.

Beer, J., R. Muschler, D. Kass and E. Somarriba. 1998. Shade management in coffee and cocoa plantations. *Agroforestry Systems* 12: 229-249.

Bernhard-Reversat, F. 1996. Nitrogen cycling in tree plantations grown on a poor sandy savanna soil in Congo. *Appl. Soil Ecol.* 4: 161-172.

Bouwman, A.F. and R. Leemans. 1995. The role of forest soils in the global carbon cycle. In W.F. McFee and F.M. Kelly (eds). Carbon Forms and Functions in Forest Soils. Soil Science Society of America, Madison, WI. pp. 503-525.

Bruenig, E.F. 1996. Conservation and Management of Tropical Rainforests: An Integrated Approach to Sustainability. *CAB International*, Wallingford, UK, 359 pp.

Buresh, R.J. and G. Tian. 1998. Soil improvement by trees in sub-Saharan Africa. *Agroforestry Systems* 38: 51-76.

Cerri, C.C., B. Volkoff and F. Andreaux. 1991. Nature and behavior of organic matter in soils under natural forest, and after deforestation, burning and cultivation near Manaus. *For. Ecol. Manage.* 38: 247-257.

Cerri, C.C., M. Bernoux and G.J. Blair. 1994. Carbon pools and fluxes in Brazilian natural and agricultural systems and the implication for the global CO_2 balance. Trans. 15th Intl. Cong. *Soil Sci., Acapulco, Mexico*, Vol. 5a: 399-406.

Cerri, C.C., M. Bernoux, D. Arrouays, B.J. Feigl and M.C. Piccolo. 2000. Carbon stocks in soils of the Brazilian Amazon. In: R. Lal, J.M. Kimble and B.A. Stewart (eds). Global Climate Change and Tropical Ecosystems. CRC Press, Boca Raton, FL. pp. 33-50.

Chander, K., S. Goyal, D.P. Nandal and K.K. Kapoor. 1998. Soil organic matter, microbial biomass and enzyme activities in a tropical agroforestry system. *Biol. Fertil. Soils* 27: 168-172.

Cuevas, E., S. Brown and J.E. Lugo. 1991. Above and below ground organic matter storage and production in a tropical pine plantation and a paired broad leaf secondary forest. *Plant Soil* 135: 257-268.

Davidson, E.A. and A.I. Hirsch. 2001. Fertile forest experiments. *Nature* 411: 431-433.

Deans, J.D., O. Diagne, D.K. Lindley, M. Dione and J.A. Parkinson. 1999. Nutrient and organic matter accumulation in *Acacia senegal* fallows over 18 years. *Forest Ecol. Manage.* 124: 153-167.

Denich, M., M. Kanashiro and P.L.G. Vlek. 2000. The potential and dynamics of carbon sequestration in traditional and modified fallow systems of the eastern Amazon region, Brazil. In: R. Lal, J.M. Kimble and B.A. Stewart (eds). Global Climate Change and Tropical Ecosystems. CRC Press, Boca Raton, FL, 229 p.

Desjardins, T., F. Andreux, B. Volkoff and C.C. Cerri. 1994. Organic carbon and ^{13}C contents in soils and soil-size fractions and their changes due to deforestation and pasture installation in eastern Amazonia. *Geoderma* 61: 103-118.

Diels, J., B. Vanlauwe, N. Sanginga, E. Coolen and R. Merckx. 2001. Temporal variations in plant delta ^{13}C values and implications for using the ^{13}C technique in long-term organic matter studies. *Soil Biol. Biochem.* 33: 1245-1251.

Dixon, R.K., S. Brown, R.A. Houghton, A.M. Solomon, M.C. Trexler and J. Wisniewski. 1994. Carbon pools and fluxes of global forest ecosystems. *Science* 263: 185-190.

Duah-Yentumi, S., R. Ronn and S. Christensen. 1998. Nutrients limiting microbial growth in a tropical forest soil of Ghana under different management. *Appl. Soil Ecol.* 8: 19-24.

Dulormne, M., J. Sierra, S.A. Sophie and F. Solvar. 1999. Capacidad de secuestración del carbono y del nitrógeno en un sistema agroforestal a base de *Gliricidia sepium* en clima tropical sub-húmedo (Guadalupe, Antillas Francesas). Unité Agropédoclimatique, Institut National de la Recherche Agromique, Domaine Duclos, Prise d´ Eau, 97170 Petit Bourg, Guadeloupe, France.

Ekanade, O., F.A. Adesina and N.E. Egbe. 1991. Sustaining tree crop production under intensive land use: An investigation into soil quality differentiation under varying cropping patterns in western Nigeria. *J. Env. Manage.* 32: 105-113.

El-Amin, E.A., I.E. Diab and S.I. Ibrahim. 2001. Influence of eucalyptus cover on some physical and chemical properties of a soil in Sudan. *Commun. Soil Sci. Plant Anal.* 32: 2267-2278.

Evans, J. 1992. Plantation Forestry in the Tropics: Tree Planting for Industrial, Social, Environmental and Agroforestry Purposes. Clarendon Press, Oxford, UK, 403 pp.

FAO. 2003. State of the World Forests. FAO, Rome, Italy.

Fearnside, P.M. 2000. Greenhouse gas emissions from land use change in Brazil's Amazon region. In: R. Lal, J.M. Kimble and B.A. Stewart (eds). Global Climate Change and Tropical Ecosystems. CRC Press, Boca Raton, FL. pp. 231-249.

Feigl, B., J. Melillo and C.C. Cerri. 1995. Changes in the origin and quality of soil organic matter after pasture introduction in Rondonia, Brazil. *Plant Soil* 175: 21-29.

Fisher, M.A., I.M. Rao, M.A. Ayarza, C.E. Lascano, J.I. Sanz, R.J. Thomas and R.R. Vera. 1994. Carbon storage by introduced deep-rooted grasses in the South American savannas. *Nature* 266: 236-238.

Fisher, R.F. 1995. Amelioration of degraded rainforest soils by plantation of native trees. *Soil Sci. Soc. Am. J.* 59: 544-549.

Fonseca, S., N.F. Barros, R.F. Novais, G.L. Leal, E.G. Loures and V. Moura-Filho. 1993. Alterações em um latossolo sob eucalipto, mata natural e pastagem: II-Propriedades orgânicas e microbiológicas. *Árvore* 17: 289-302.

Ghuman, B.S. and R. Lal. 1990. Nutrient addition into soil by leaves of *Cassia siamea* and *Gliricidia sepium* grown on an Ultisol in southern Nigeria. *Agroforestry Systems* 10: 131-133.

Grainger, A. 1988. Estimating areas of degraded tropical lands requiring replenishment of forest cover. *Int. Tree Crops J.* 5: 31-61.

Haggar, J.P., E.V.J. Tanner, J.W. Beer and D.C.L. Kass. 1993. Nitrogen dynamics of tropical agroforestry and annual cropping systems. *Soil Biol. Biochem.* 25: 1363-1378.

Hahn, S.K., E.R. Terry, K. Leuschner, I.O. Akobundu, S. Okali and R. Lal. 1979. Cassava improvement in Africa. *Field Crops Res.* 2: 193-226.

Härdter, R., W.Y. Chow and O.S. Hock. 1997. Intensive plantation cropping, a source of sustainable food and energy production in the tropical rainforest areas in southeast Asia. *Forest Ecol. Manage.* 93: 93-102.

Hartemink, A.E. 2003. Soil Fertility Decline in the Tropics with Case Studies on Plantations. CABI Publishing, Wallingford, UK. 360 pp.

IPCC 2001. Climate Change 2001: The Scientific Basis. Oxford Univ. Press, Oxford, UK.

Jenkinson, D.S. and A. Ayanaba. 1977. Decomposition of carbon-14 labeled plant material under tropical conditions. *Soil Sci. Soc. Am. J.* 41: 912-915.

Jin, J.Y., L. Bao and W.L. Zhang. 1999. Improving nutrient management for sustainable development of agriculture in China. In: E.M.A. Smaling, O. Oenema and L.O. Fresco (eds). Nutrient Disequilibria in Agroecosystems: Concepts and Case Studies. CAB International, Wallingford, UK. pp. 174-157.

Johnson, D.W. 1992. Effects of forest management on soil carbon storage. *Water, Air and Soil Pollution* 64: 83-120.

Juo, A.S.R., K. Franzluebbers, A. Dabiri and B. Ikhile. 1995. Changes in soil properties during long-term fallow and continuous cultivation after forest clearing in Nigeria. *Agric. Ecosyst. Env.* 56: 9-18.

Kang, B.T. 1987. Nitrogen cycling in multiple cropping system. In: J.R. Wilson (ed). Advances in Nitrogen Cycling in Agricultural Ecosystems. CAB Intl., Wallingford, UK. pp. 333-348.

Kang, B.T. 1997. Alley cropping-soil productivity and nutrient cycling. *Forest Ecol. Manage.* 91: 75-82.

Kass, D.C., A. Barrantes, W. Bermudez, W. Campos, M. Jiménez and J. Sanchez. 1989. Resultados de seis años de investigación de cultivo en callejones (alley cropping), en la Montaña, Turrialba, Costa Rica. *El Chasqui Turrialba* 19: 5-24.

Kass, D.C., R. Sylvester-Bradley and P. Nygren. 1997. The role of nitrogen fixation and nutrient supply in some agroforestry systems of the Americas. *Soil Biol. Biochem.* 29: 775-785.

Kumar, B.M. and J.K. Deepu. 1992. Litter production and decomposition dynamics in moist deciduous forests of the western Ghats in Penninsular India. *For. Ecol. Manage.* 50: 181-201.

Kunu, T. and A.E. Hartemink. 1997. Soil chemical properties under primary forest and coffee in the Kutubu area of Papua, New Guinea. *Papua New Guinea J. of Agric., For. & Fisheries* 40: 1-5.

Lal, R. 1975. Role of mulching techniques in tropical soils and water management. IITA Tech. Bull. 1, 38 pp.

Lal, R. 1989a. Agroforestry systems and soil surface management of a tropical Alfisol. I. Soil erosion and nutrient loss. *Agroforestry Systems* 8: 97-111.

Lal, R. 1989b. Agroforestry systems and soil surface management of a tropical Alfisol. II. Changes in soil chemical properties. *Agroforestry Systems* 8: 113-132.

Lal, R. 1989c. Agroforestry systems and soil surface management of a tropical Alfisol. III. Effects on soil physical and mechanical properties. *Agroforestry Systems* 8: 197-215.

Lasco, R.D. 1991. Herbage decomposition of some agroforestry species and their effects as mulch on soil properties and crop yield. Philippines Univ., Los Baños, Laguna, Philippines: 170 p.

Leaky, R.B. and A.C. Newton. 1994. Domestication of Tropical Trees for Timber and Non-Timber Products. UNESCO, Paris.

Lepsch, I.F. 1980. Influência do cultivo de *Eucalyptus* e *Pinus* nas propriedades químicas de solos sob Cerrado. *Rev. Bras. Ciência Solo* 4: 103-107.

Lugo, A.E. and M.J. Sanchez. 1986. Land use and organic carbon content of some sub-tropical soils. *Plant Soil* 96: 185-196.

Luizão, R.C., T.A. Bonde and T. Rosswall. 1992. Seasonal variation of soil microbial biomass-the effect of clear-felling tropical rainforest and establishment of pasture in the central Amazon. *Soil Biol. Biochem.* 24: 805-813.

Lumbanraja, J., T. Syam, H. Nishide, A.K. Mahi, M. Utomo and M. Kimura. 1998. Deterioration of soil fertility by land use changes in south Sumatra, Indonesia–from 1970 to 1990. *Hydrological Processes* 12: 2003-2013.

Lupwayi, N.Z. and I. Haque. 1999a. Leucaena hedgerow intercropping and cattle manure application in the Ethiopian highlands. I. Decomposition and nutrient release. *Biol. Fertil. Soils* 28: 182-195.

Lupwayi, N.Z. and I. Haque. 1999b. Leucaena hedgerow intercropping and cattle manure application in the Ethiopian highlands. III. Nutrient balance. *Biol. Fertil. Soils* 28: 204-211.

Lupwayi, N.Z., I. Haque, A.R. Saka and D.E.K.A. Siaw. 1999. Leucaena hedgerow intercropping and cattle manure application in the Ethiopian highlands. II. Maize yields and nutrient uptake. *Biol. Fertil. Soils* 28: 196-203.

Madeira, M.A.V., P.P. Andreux and J.M. Portal. 1989. Changes in soil organic matter characteristics due to reforestation with *Eucalyptus globulus*, in Portugal. *Sci. Total Environ.* 81/82: 481-488.

Malab, S.C., J.I. Rosario, E.O. Agustin, F.C. Pastor and S.M. Pablico. 1997. *Desmanthus* as hedgerows in agroforestry cropping systems. *Philippine J. Crop Sci.* 22: 26 p.

Manlay, R.J., M. Kaire, D. Masse, J.L. Chotte, G. Ciornei and C. Floret. 2002. Carbon, nitrogen and phosphorus allocation in Agroecosystems of a west African savanna. I. The plant component under semi-permanent cultivation. *Agric. Ecosyst. Environ.* 88: 215-232.

Margulis, S. 2004. Causes of Deforestation of the Brazilian Amazon. World Bank Working Paper #22, Washington, DC. 77 pp.

Maroko, J.B., R.J. Buresh and P.C. Smithson. 1998. Soil nitrogen availability as affected by fallow-maize systems on two soils in Kenya. *Biol. Fertil. Soils* 26: 229-234.

Mazzarino, M.J., L. Szott and M. Jiménez. 1993. Dynamics of soil total C and N, microbial biomass, and water soluble C in tropical Agroecosystems. *Soil Biol. Biochem.* 25: 205-214.

Montagnini, F. 2000. Accumulation in above-ground biomass and soil storage of mineral nutrients in pure and mixed plantations in a humid tropical lowland. *For. Ecol. Manage.* 134: 257-270.

Montagnini, F. and F. Sancho. 1990. Impacts of native trees on tropical soils: A study in the Atlantic lowlands of Costa Rica. *Ambio* 19(8): 386-390.

Montagnini, F. and C. Porras. 1998. Evaluating the role of plantations as carbon sinks: An example of an integrative approach from the humid tropics. *Env. Manage.* 22: 459-470.

Moraes, J.F.L., C.C. Cerri, J.M. Melillo, D. Kicklighter, C. Neill, D.L. Skole and P.A. Steudler. 1995. Soil carbon stock of the Brazilian Amazon. *Soil Sci. Soc. Am. J.* 59: 244-247.

Mugendi, D.N., P.K.R. Nair, J.N. Mugwe, M.K. O'Neill and P. Woomer. 1999. Alley cropping of maize with *Calliandra* and *Leucaena* in the sub-humid highlands of Kenya. Part I. Soil fertility changes and maize yield. *Agrofor. Syst.* 46: 39-50.

Mulongoy, K. and M.K. van der Meersch. 1988. Nitrogen contribution by *Leucaena leucocephala* prunings to maize in an alley cropping system. *Biol. Fert. Soils* 6: 282-285.

Natural Research Council. 1993. Sustainable Agriculture and Environment in the Humid Tropics. National Academy Press, Washington, DC.

Neill, C., P.A. Steudler, J.M. Melillo, C.C. Cerri, J.F.L. Moraes, M.C. Piccolo and M. Brito. 1997. Carbon and nitrogen stocks following forest clearing for pasture in the southwestern Brazilian Amazon. *Ecol. Appl.* 7: 1216-1225.

Neill, C. and E.A. Davidson. 2000. Soil carbon accumulation or loss following deforestation for pasture in the Brazilian Amazon. In: R. Lal, J.M. Kimble and B.A. Stewart (eds). Global Climate Change and Tropical Ecosystems. CRC Press, Boca Raton, FL. pp. 211.

Nykvist, N., H. Grip, S.B. Sim, A. Malmer and F.K. Wong. 1994. Nutrient losses in forest plantations in Sabah, Malaysia. *Ambio* 23: 210-215.

Oades, J.M., G.P. Gillman and G. Uehara. 1989. Interaction of soil organic matter and variable-charge clays. In: D.C. Coleman, J.M. Oades and G. Uehara (eds). Dynamics of Soil Organic Matter in Tropical Ecosystems. Niftal Project, Univ. of Hawaii, Honolulu, HI. pp. 69-95.

Ogunkunle, A.O. and O.O. Eghaghara. 1992. Influence of land use on soil properties in a forest region of southern Nigeria. *Soil Use and Manage.* 8: 121-125.

Pacala, S. and R. Socolow. 2004. Stabilization wedges: Solving the climate problem for the next 50 years with current technologies. *Science* 305: 968-972.

Parrotta, J.A. 1992. The role of plantation forests in rehabilitating degraded tropical ecosystems. *Agric., Ecosystems Environ.* 41: 115-133.

Patma-Vityakon, V. 1991. Relationships between organic matter and some chemical properties of sandy soils with different land use and soil management. *Thai J. Soil and Fert.* 13: 254-264.

Prentice, I.C. 2001. The carbon cycle and atmospheric carbon dioxide. Climate Change 2001: The Scientific Basis. IPCC, Cambridge Univ. Press, Cambridge, UK. pp. 183-237.

Prosser, I.P., K.J. Hailes, M.D. Melville, R.P. Avery and C.J. Slade. 1993. A comparison of soil acidification under *Eucalyptus* forest and unimproved pasture. *Aust. J. Soil Res*. 31: 245-254.

Pushparajah, E. 1998. The oil palm–a very environmentally friendly crop. *The Planter* (Kuala Lumpur) 74: 63-72.

Resck, D.V.S., C.A. Vasconcellos, L. Vilela and M.C.M. Macedo. 2000. Impact of conversion of Brazilian Cerrados to cropland and pasture land on soil C pool and dynamics. In: R. Lal, J.M. Kimble and B.A. Stewart (eds). Global Climate Change and Tropical Ecosystems. CRC Press, Boca Raton, FL. pp. 169-196.

Resh, S.C., D. Binkley and J.A. Parrotta. 2002. Greater soil carbon sequestration under nitrogen-fixing trees compared with Eucalyptus species. *Ecosystems* 5: 217-231.

Roose, E. and B. Barthes. 2001. Organic matter management for soil conservation and productivity restoration in Africa: A contribution from Francophone research. *Nutr. Cycle Agroeco*. 61: 159-170.

Ruppenthal, M., D.E. Leihner, N. Steinmüller and M.A. El Sharkawy. 1997. Losses of organic matter and nutrients by water erosion in cassava-based cropping systems. *Expl. Agric*. 33: 487-498.

Salako, F.K., O. Babalola, S. Hauser and B.T. Kang. 1999. Soil macroaggregate stability under different fallow management systems and cropping intensities in southwestern Nigeria. *Geoderma* 91: 103-123.

Salako, F.K. and G. Tian. 2001. Litter and biomass production from planted and natural fallows on a degraded soil in southwestern Nigeria. *Agroforestry Systems* 51: 239-251.

Sanchez, P.A. 1995. Science in agroforestry. *Agroforestry Systems* 30: 5-55.

Sanchez, P.A., M.P. Gichuru and L.B. Katz. 1982. Organic matter in major soils of the tropical and temperate regions. Trans. 12th Intl. Cong. Soil Sci. 1 (New Delhi).

Sanchez, P.A., C.A. Palm, C.B. Davey, L.T. Szott and C.E. Russell. 1985. Tree crops as soil improvers in the humid tropics. In: M.G.R. Cannell and J.E. Jackson (eds). Attributes of Trees as Crop Plants. Institute of Terrestrial Ecology, Huntington, UK. pp. 79-124.

Sanchez, P.A., P.L. Woomer and C.A. Palm. 1994. Agroforestry approaches for rehabilitating degraded lands after tropical deforestation. JIRCAS Intl. Symp. Series Vol. 1, Tsukuba, Japan.

Sanginga, N. and M.J. Swift. 1992. Nutritional effects of *Eucalyptus* litter on the growth of maize (*Zea mays*). *Agric. Ecosyst. Environ*. 41: 55-65.

Sarmiento, L. and P. Bottner. 2002. Carbon and nitrogen dynamics in soils with different fallow times in the high tropical Andes: Indications for fertility restoration. *Appl. Soil Ecol*. 19: 79-89.

Sauerbeck, D.R. 2001. CO_2 emissions and C sequestration by agriculture-perspectives and limitations. *Nutr. Cycles Agroeco*. 60: 253-266.

Schlesinger, W.H. and J. Lichter. 2001. Limited carbon storage in soils and litter of experimental forest plots under increased atmospheric CO_2. *Nature* 411: 466-469.

Schroeder, P. 1992. Carbon storage potential of short rotation tropical tree plantations. *For. Ecol. Manage*. 50: 31-41.

Schroth, G., N. Poiday, T. Morshauser and W. Zech. 1995. Effects of different methods of soil tillage and biomass application on crop yields and soil properties in agroforestry with high tree competition. *Agric. Ecosyst. Environ*. 52: 129-140.

Schroth, G., S.A. D´Angelo, W.G. Teixeira, D. Haag and R. Lieberei. 2002. Conversion of secondary forest into agroforestry and monoculture plantations in Amazonia: Consequences for biomass, litter and soil carbon stocks after 7 years. *For. Ecol. Manage.* 163: 131-150.

Smith, C.K., F. de Asis Oliveira, H.L. Gholz and A. Baima. 2002. Soil carbon stocks after forest conversion to tree plantations in lowland Amazonia, Brazil. *For. Ecol. Manage.* 164: 257-263.

Stanley, W. and F. Montagnini. 1999. Biomass and nutrient accumulation in pure and mixed plantations of indigenous tree species grown on poor soils in the humid tropics of Costa Rica. *For. Ecol. Manage.* 113: 91-103.

Tornquist, C.G., F.M. Hons, S.E. Feagley and J. Haggar. 1999. Agroforestry system effects on soil characteristics of the Sarapiqui region of Costa Rica. *Agric. Ecosyst. Environ.* 73: 19-28.

Trumbore, S.E., E.A. Davidson, P.B. de Camargo, D.C. Nepstad and L.A. Martinelli. 1995. Below-ground cycling of C in forests and pastures of eastern Amazonia. *Global Biogeochem. Cycles* 9: 512-528.

Turner, J. and M. Lambert. 2000. Change in organic carbon in forest plantation soils in eastern Australia. *For. Ecol. Manage.* 133: 231-247.

Van Noordwijk, M., C. Cerri, P.L. Woomer, K. Nugroho and M. Bernoux. 1997. Soil carbon dynamics in the humid tropical forest zone. *Geoderma* 79: 187-225.

Veldkamp, E. 1994. Organic carbon turnover in three tropical soils under pasture after deforestation. *Soil Sci. Soc. Am. J.* 58: 175-180.

Wada, K. and S. Aomine. 1973. Soil development during the Quaternary. *Soil Sci.* 116: 170-177.

Wada, K. 1985. The disruptive properties of Andosols. In: B.A. Stewart (ed). Advances in Soil Science, Vol. 2, Springer Verlag, NY. pp. 173-229.

Wada, K. (Ed.). 1986. Andosols in Japan. Kyushu Univ. Press, Fukuoka, Japan.

Woomer, P.L. 1993. Modelling soil organic matter dynamics in tropical ecosystems: Model adoption, uses and limitations. In: K. Mulongoy and R. Merckk (eds). Soil Organic Matter and Sustainability of Tropical Agriculture. J. Wiley & Sons, NY. pp. 279-294.

Woomer, P.L., A. Martin, A. Albrecht, D.V.S. Resck and H.W. Scharpenseel. 1994. The importance and management of soil organic matter in the tropics. In: P.L. Woomer and M.J. Swift (eds). The Biological Management of Tropical Soils. J. Wiley & Sons, Chichester, UK. pp. 47-80.

Woomer, P.L., C.A. Palm, J. Alegre, C. Castillo, D.G. Cordeiro, K. Hairach, J. Dotto-Same, A. Moukam, A. Riese, V. Rodrigues and M. van Noordwijk. 2000. Slash and burn effects on carbon stocks in the humid tropics. In: R. Lal, J.M. Kimble and B.A. Stewart (eds). CRC Press, Boca Raton, FL. pp. 99-115.

Wright, D.G., R.W. Mullen, W.E. Thomason and W.R. Raun. 2001. Estimated land area increase of agricultural ecosystems to sequester excess atmospheric carbon dioxide. *Commun. Soil Sci. Plant Anal.* 32: 1803-1812.

Zinn, Y.L., D.V.S. Resck and J.E. da Silva. 2002. Soil organic carbon as affected by afforestation with *Eucalyptus* and *Pinus* in the Cerrado region of Brazil. *For. Ecol. Manage.* 166: 285-294.

Carbon Sequestration in Pastures, Silvo-Pastoral Systems and Forests in Four Regions of the Latin American Tropics

María Cristina Amézquita
Muhammad Ibrahim
Tangaxuhan Llanderal
Peter Buurman
Edgar Amézquita

SUMMARY. Tropical America (TA) holds 8% of the world's population, 11% of the world's continental area, 23% and 22%, respectively, of the world's forest and water resources, and 13% of the world's pasture and agro-pastoral land, this representing 77% of TA's agricultural land. Recent interest in carbon sequestration and preliminary research suggest

María Cristina Amézquita is Scientific Director, Carbon Sequestration Project, The Netherlands Cooperation CO-010402, CIAT's Science Park, Cali, Colombia (E-mail: m.amezquita@cgiar.org).

Muhammad Ibrahim is Agroforestry Scientist, CATIE, Turrialba, Costa Rica (E-mail: mibrahim@catie.ac.cr).

Tangaxuhan Llanderal is a PhD student of Agroforestry, CATIE, Turrialba, Costa Rica (E-mail: tllanderal@catie.ac.cr).

Peter Buurman is Associate Professor and Head of Soil Science and Geology, Department of Environmental Sciences, Wageningen University, The Netherlands (E-mail: peter.buurman@wur.nl).

Edgar Amézquita is Soil Scientist, CIAT, Cali, Colombia (E-mail: e.amezquita@cgiar.org).

[Haworth co-indexing entry note]: "Carbon Sequestration in Pastures, Silvo-Pastoral Systems and Forests in Four Regions of the Latin American Tropics." Amézquita, María Cristina et al. Co-published simultaneously in *Journal of Sustainable Forestry* (Food Products Press, an imprint of The Haworth Press, Inc.) Vol. 21, No. 1, 2005, pp. 31-49; and: *Environmental Services of Agroforestry Systems* (ed: Florencia Montagnini) Food Products Press, an imprint of The Haworth Press, Inc., 2005, pp. 31-49. Single or multiple copies of this article are available for a fee from The Haworth Document Delivery Service [1-800-HAWORTH, 9:00 a.m. - 5:00 p.m. (EST). E-mail address: docdelivery@haworthpress.com].

Available online at http://www.haworthpress.com/web/JSF

doi:10.1300/J091v21n01_02

that well-managed pasture systems in TA could provide a good combination of economic production, poverty reduction, recovery of degraded areas and delivery of environmental services, particularly, carbon sequestration. This paper presents 3-year research results generated by the "Carbon Sequestration Project, The Netherlands Cooperation CO-01002" on soil carbon stocks (SCS) for a range of pasture and silvo-pastoral systems prevalent in agro-ecosystems of TA compared to native forest and degraded land. In the tropical Andean hillsides, Colombia (1350-1900 m.a.s.l., 1800 mm rainfall/yr, 14-18°C mean annual temperature, medium to high slopes, medium fertility soils), SCS from *Brachiaria decumbens* pastures were statistically lower than those from native forest, but higher than those from natural regeneration of a degraded pasture (fallow land), degraded pasture and mixed-forage bank. In contrast, in the humid tropical forest of the Atlantic Coast, Costa Rica (200 m.a.s.l., 28-35°C, 3500 mm/year, poor acid soils), pasture or silvo-pastoral systems with native or planted pasture species such as *Ischaemum ciliare*, *Brachiaria brizantha* + *Arachis pintoi* and *Acacia mangium* + *Arachis pintoi* showed statistically higher SCS than native forest. Similar rankings were found in the humid tropical forest of Amazonia, Colombia (800 m.a.s.l., 30-42°C, 4200 mm/yr, flat, very poor acid soils) where improved *Brachiaria* pastures (monoculture and legume-associated) showed statistically higher SCS than native forest. In the sub-humid tropical forest of the Pacific Coast, Costa Rica (200 m.a.s.l., 6-month dry season, 2200 mm/year, poor acid soils) no statistical differences in SCS were found between land-use systems. In tropical ecosystems, improved pasture and silvo-pastoral systems show comparable or even higher SCS than those from native forests, depending on climatic and environmental conditions (altitude, temperature, precipitation, topography and soil), and represent attractive alternatives as C-improved systems.

KEYWORDS. Tropics, soil carbon stocks, tropical pastures, silvo-pastoral systems

INTRODUCTION

Tropical America (TA) comprises Mexico, Central America, the Caribbean and South America, excluding Argentina, Chile and Uruguay. It covers 1,688 million ha representing 11% of the world continental area, and houses 432 million people representing 8% of the total world population. Forests cover 41% of its territory–a very high proportion compared with the world proportion in forests (28%)–and represent 22% of the world forest area (FAO, 2000). Water renewable resources in the region are abundant, representing 22% of the world freshwater resources, with an average per-capita availability of 35,405 m^3, almost five times the corresponding world average of 7,176 m^3. TA's agricultural land amounts to 548 million ha corresponding to 32% of its territory and to 11% of the world agricultural land. Of its 432 million inhabitants, 100 million (23%) are farmers, representing 4% of the world population who live from agricultural and livestock production activities. TA's agricultural land includes crop and pasture (both native and introduced), silvo-pastoral and agro-silvo-pastoral land (FAO, 2000).

Pasture, silvo-pastoral and agro-silvo-pastoral land represent 77% of TA's total agricultural land, mostly on poor acid soils, while crops cover the remaining 23%, mostly located on better quality soils, with a pastures/crop land ratio of 3.4, higher than the world ratio of 2.3. TA holds 21% of the world's cattle inventory and 18% of the world's inventory of dairy cattle. Pasture systems, as well as meat and milk production, are concentrated in four countries: Brazil, Mexico, Colombia and Venezuela. Together they hold 76% of the pasture and agro-silvo-pastoral land, 84% of total cattle inventory, 85% and 83%, respectively, of meat and milk production of TA (Vera et al., 1993). The major tropical ecosystems where meat and milk are produced are: savannah (250 million ha, 243 of them in the above-mentioned countries), tropical forest (with near 44 million ha, almost all in the same four countries (Vera et al., 1993), and Andean hillsides (96 million ha, in Colombia, Ecuador, Perú and Venezuela). This area excludes the hillsides of Brazil, the hillsides of Central America, and the foothills of the Andes in Brazil, extensive areas that are essentially identical in their topography and land use (Jones, 1993; Amézquita et al., 1998).

Meat and milk production in TA have important socio-economic significance. Its consumption is high in most cities, meat and milk representing, respectively, 12-26% and 7-13% of total family expenditure (Rubinstein and Nores, 1980). Meat consumption per capita/year in TA ranges from 7 kg to 38 kg, while in other tropical regions of the world it ranges from 0.7 kg to 2.6 kg in Southeast Asia and from 3.6 kg to 9.6 kg in tropical Africa (Valdés and Nores, 1979). The positive impact of improved and well-managed pasture systems on productivity, farmers' socio-economic condition, export market competitiveness and economic development of TA's countries has been amply documented (CIAT, 1976-2000; Rubinstein and Nores, 1980; Sanint et al., 1984; Toledo, 1985; Rivas et al., 1998; Vera et al., 1993).

This paper presents 3-year research results generated by the "Carbon Sequestration Project, The Netherlands Cooperation CO-01002" on soil carbon stocks (SCS) for a range of pasture and silvo-pastoral systems prevalent in agro-ecosystems of TA compared to native forest (positive control) and degraded land (negative control).

Land-Use Change in Latin America

Conversion of forests to crops and pastures has been the most important land-use change in TA in the last fifty years (Kaimowitz, 1996). In Colombia (Table 1) the decline in forest area has been accompanied by an increase in pasture area with no major change in cropped area, as pastures are established once soils are too degraded for crop production. However, cattle production systems are not the only cause of deforestation (Kaimowitz, 1996). Deforestation is attributed to population pressure, national policies, ease of access to the forests, high initial soil fertility with favorable conditions for crop and pasture establishment, and production and marketing interests of multinational companies searching for highly profitable timber production (Browder, 1988; Sader and Joyce, 1988; Veldkamp, 1993; 1994). After deforestation and crop and pasture establishment, many areas have been abandoned due to production decline caused by mismanagement, which has caused degradation in more than 60% of the TA's pasture area (CIAT, 1999).

TABLE 1. Land Use Change in Colombia 1950-2000. Area (million ha, %).

Land Use	1950 area, (%)	1970 area, (%)	1978 area, (%)	1987 area, (%)	1995 area, (%)	2000 area, (%)
Crops	5.0 (4)	7.6 (7)	8.8 (8)	5.3 (5)	4.4 (4)	5.0 (5)
Pastures	14.6 (13)	17.5 (15)	20.5 (18)	40.1 (35)	35.5 (31)	45.0 (39)
Forest	94.6 (83)	89.1 (78)	84.9 (74)	68.7 (60)	74.2 (65)	64.2 (56)
Total	114.2 (100)	114.2 (100)	114.2 (100)	114.2 (100)	114.2 (100)	114.2 (100)

Adapted from Ramírez C. and Ortiz R., 1997.

Carbon Sequestration and Land Use

Recently, there has been an increased interest in carbon sequestration worldwide. The Kyoto Protocol (Conference of the Parties in its third session, COP3, 1997) and subsequent agreements of the United Nations (COP4 to COP9, 1998-2003) considered reforestation and afforestation to be land-use systems suitable for economic incentives in developing countries through Clean Development Mechanism (CDM) and international carbon trading. Although environmental considerations might suggest that partial reforestation of areas currently under pasture could potentially contribute to carbon sequestration, this would cause a serious threat to the socio-economic welfare of farmers who derive their living from livestock activities, and to food availability (especially milk and meat) for the population. Therefore, the combination of agricultural production and environmental services (particularly carbon sequestration) through improved and well-managed pasture, agro-pastoral and silvo-pastoral systems, appears to be a good alternative. Additionally, the approval for European Union countries to contribute to their greenhouse gas (GHG) emission reduction through carbon sequestration in grassland systems (Marrakech Accords, COP7, 2001), and the United States' motivation to provide farmer's incentives for carbon sequestration in grasslands (United States Department of Agriculture, June 2003) makes this alternative particularly attractive to countries in TA. Tropical pasture, agro-pastoral and silvo-pastoral systems could be also considered to be land-use systems suitable for international carbon trading and other economic incentives for developing countries.

OBJECTIVES OF THE PRESENT RESEARCH

This paper presents three-year research results on the evaluation of soil carbon stocks (SCS) in long-established pasture and silvo-pastoral systems (10-16 years under commercial production), native forests and degraded land, in four regions of TA: tropical Andean hillsides (Colombia), sub-humid tropical forest (Costa Rica), humid tropical forest (Costa Rica), and humid tropical forest, Amazonia (Colombia), with a possible addition from 2005 onwards of the savannah ecosystem, with research sites in Colombia (Amézquita, 2002).

Results presented in this article correspond to the first C-sampling cycle of long-established systems in all ecosystems under investigation, with fieldwork conducted in 2002 and 2003. A second C-sampling cycle of all long-established systems in all ecosystems is being carried out at present and is envisaged to continue throughout 2005, with research results expected in 2006. Long-established systems under evaluation, due to their age, are thought to have reached equilibrium in their soil carbon accumulation capacity. Therefore, C-sampling cycles are considered 'replications in time' implemented to obtain more precise estimates of SCS, although significant differences in systems' SCS would not be expected between both C-samplings. This hypothesis will be the subject of a future research publication.

Results presented in this article are part of the research agenda of the long-term international project entitled "Research Network for the Evaluation of Carbon Sequestration Capacity of Pasture, Agro-pastoral and Silvo-pastoral Systems in the American Tropical Forest Ecosystem." The project aims at identifying pasture, agro-pastoral and silvo-pastoral systems that represent attractive economic alternatives to the farmer and show high levels of carbon sequestration and carbon stocks, comparing them with two reference states: degraded land, or degraded pasture (negative control), and native forest (positive control). In addition, it aims at providing recommendations to policy makers at local, national and international levels about appropriate pasture systems, while also considering their socio-economic benefit and provision of environmental services, particularly carbon sequestration.

This paper presents results on SCS from a range of improved pasture and silvo-pastoral systems evaluated in farms' networks located in the four regions mentioned above. An 'improved' pasture or silvo-pastoral system has one or more of the following characteristics: appropriate soil management, use of pasture or tree species that are more productive than local varieties, and use of appropriate plant management practices.

METHODOLOGY

Land-Use Systems and Sites for Evaluation of Soil Carbon Stocks

SCS were evaluated in long-established pasture and silvo-pastoral systems under grazing on commercial farms, 10-16 years after establishment depending on the ecosystem. Table 2 describes the range of land use systems evaluated at each region. Field research was conducted in three networks of farms located in the project regions. The

TABLE 2. Land-Use Systems Evaluated per Ecosystem.

Tropical Andean Hillsides	
Zone 1 (Dovio, Colombia)	**Zone 2 (Dagua, Colombia)**
1. Native Forest (positive control)	1. Native Forest (positive control)
2. Improved pasture: *Brachiaria decumbens* under grazing	2. Improved pasture: *Brachiaria decumbens* under grazing
3. Mixed Forage Bank for "cut and carrying," composed of 5 tree species: *Trichanthera gigantea, Morus* spp., *Erythrina edulis, Boehmeria nivea,* and *Tithonia diversifolia*	3. Mixed Forage Bank for "cut and carrying," composed of 4 species: *Trichanthera gigantea, Morus* spp., *Erythrina fusca* and *Tithonia diversifolia*
4. Degraded pasture (negative control): King grass (*Pennisetum purpureum* and *Pennisetum benthami*) + *Melinis minutiflora* pasture, degraded, overgrazed, with presence of *Psidium guajaba* tress and weed invasion (*Pteridium aquilinum* and *Sida* spp.)	4. Natural regeneration of a degraded pasture: *Hyparrhenia rufa* pasture, non-grazed, invaded by weeds such as *Andropogon bicornis, Andropogon leucostachyus, Pteridium aquilinum, Baccharis trinervis* and *Calea* spp.
	5. Secondary Forest
	6. Degraded land (negative control)
Humid and sub-humid tropical forest, Costa Rica	
Zone 1 (Pocora, Atlantic coast, humid tropical forest)	**Zone 2 (Esparza, Pacific coast, sub-humid tropical forest)**
1. Native Forest (positive control)	1. Native Forest (positive control)
2. Improved silvo-pastoral system: *Acacia mangium* + *Arachis pintoi*	2. Improved silvo-pastoral system: *Brachiaria brizantha* + *Cordia alliodora* + *Guazuma ulmifolia*
3. Improved pasture system: *Brachiaria brizantha* + *Arachis pintoi*	3. Improved pasture: *Brachiaria decumbens* under grazing
4. Improved pasture: *Brachiaria brizantha*	4. Native grass: *Hyparrhenia rufa*
5. Native grass: *Ischaemum ciliare*	5. Forage Bank: *Cratylia argentea*
6. Degraded pasture: *Ischaemum ciliare,* overgrazed (negative control)	6. Secondary Forest
	7. Degraded Pasture: *Hyparrhenia rufa,* overgrazed (negative control)
Humid tropical forest, Amazonia, Colombia	
Zone 1("La Guajira" farm, flat topography, Florencia, Colombia)	**Zone 2 ("Pekin" farm, mild-slope topography, Florencia, Colombia)**
1. Native Forest (positive control)	1. Native Forest (positive control)
2. *Brachiaria decumbens* + native legumes	2. *Brachiaria decumbens* + native legumes
3. *Brachiaria humidicola* + native legumes	3. *Brachiaria humidicola* + native legumes
4. *Brachiaria decumbens*	4. *Brachiaria decumbens*
5. *Brachiaria humidicola*	5. *Brachiaria humidicola*
6. Degraded pasture: overgrazed *Brachiaria decumbens,* invaded by weeds	6. Degraded pasture: overgrazed *Brachiaria* + *Desmodium ovalifolium* + *Calopogonium mucuniodes* pasture, invaded by weeds

tropical Andean hillsides farmers' network consisted of six small farms (2-12 ha) of dual-purpose cattle production, located in two sites of the tropical Andean hillsides: Dovio, Colombia (1900 m.a.s.l., 1800 mm/yr, 14°C, high slopes, pH 5.5-6.5, medium fertility soils) and Dagua, Colombia (1350 m.a.s.l., 1800 mm/yr, 20°C, medium slopes, pH 5.0-6.1, less fertile soils). The sub-humid and humid tropical forest of Costa Rica farmers' networks consisted of eight small and medium-size farms (7-92 ha), four of them located in the sub-humid tropical forest of the Pacific Coast near Esparza, Costa Rica (200 m.a.s.l., 28°C, six months of dry season and six months with 2200 mm, poor acid soils) and four in the humid tropical forest, near Pocora, Costa Rica (200 m.a.s.l., 28-35°C, 3500 mm/year, poor acid soils). The Colombian Amazonia humid tropical forest farmers' network consisted of four medium-size farms (250-500 ha) located near Florencia, Colombia (500-800 m.a.s.l., 30-42°C, 4200 mm/yr, flat or mild-slope area with slopes < 10%, pH 4.2-4.6, very poor acid soils).

Soil Sampling Design for Evaluation of Soil Carbon Stocks

A soil sampling design controlling factors affecting SCS (site conditions, slope or main gradient, land-use system, and soil depth) was used. SCS were evaluated at four soil depths (0-10, 10-20, 20-40 and 40-100 cm) using two to three replications per system and twelve sampling points per system/replication. All samples were composite samples. Total C, oxidizable C, total N, P, CEC, pH, soil texture and bulk density were evaluated at each soil pit and depth. Total C, oxidizable C and stable C (the latter expressed as the difference between total C and oxidizable C) were expressed in Megagrams (tons) C/ha/depth for each soil depth. Oxidizable carbon was determined by wet oxidation according to the Walkley-Black technique (USDA, 1996). Total carbon was determined by dry combustion at 120°C. CEC at pH 7 was determined by the ammonium acetate method (USDA, 1996). The pH in water was determined with a 1:5 solid:solution ratio (USDA, 1996). Soil texture was determined by pipette method and sieving (USDA, 1996). Available P was determined using the Bray-I method (USDA, 1996).

Statistical Analysis

For statistical comparisons of SCS between land-use systems, calculations based on fixed soil mass according to Ellert et al. (2002)–which adjusts SCS to a constant soil weight value per sampling point for a given soil depth–without subdivision in soil horizons as modified by Buurman et al. (2004), were carried out for total C and stable C, using ANOVA models consistent with the sampling design. Following Buurman et al. (2004), the minimum soil mass per sampling point to 1 m depth was used as reference for each experimental area. Although the fixed soil mass method is more accurate, many authors conveniently use fixed depth soil carbon stock estimates instead. We therefore present both estimates for our research sites. A perfect correlation between fixed soil mass and fixed soil depth estimates should be expected exclusively when bulk densities do not show a large variation with depth. The two calculation methods–fixed soil depth and fixed soil mass–were compared in terms of absolute SCS estimates. Multivariate statistical analysis techniques, such as Principal Component Analysis and Cluster Analysis, were used to identify relationships between SCS and soil parameters and to group sampling points that were similar in their soil conditions and their level of SCS.

Initial Soil Characterization

A complete soil characterization of each land-use system at the four ecosystems considered was performed at the beginning of the project. All soil parameters described earlier were evaluated at four soil depths: 0-10 cm, 10-20 cm, 20-40 cm, and 40-100 cm (Amézquita, 2003; Amézquita et al., 2003).

Land-Use History

Based on verbal information given by landowners, a 50-year land-use history was recorded for each land use system evaluated. Table 3 shows this information for tropical Andean hillsides land-use systems. Similar information was recorded for land-use systems from the sub-humid and humid tropical forest ecosystems of Costa Rica and the humid tropical forest of Amazonia, Colombia (Amézquita et al., 2003; Amézquita et al., 2004).

Socio-Economic Evaluation

Socio-economic characterization of improved and conventional farms within each ecosystem was carried out through participatory workshops, field days and socio-economic surveys of farmers. The main pur-

TABLE 3. Land Use Change (last 50 years). Tropical Andean Hillsides.

Area 1: Dovio (1900 m.a.s.l.), Colombia

Present Land Use System	Initial Land Use	1950's	1960's	1970-1977	1977-1986	1986-1988	1988	1988-2002
Degraded Pasture	Forest	Sugar cane	Abandoned land	Fruit trees (Tomate de árbol)	Pasture (*Melinis minutiflora*)	King grass var Taiwan	Degraded King grass + trees + maize + pineapple	Degraded King grass Pasture
Improved Pasture	Forest	Sugar cane	Coffee + Guamo	Fruit trees (Tomate de árbol)	Abandoned land		*Brachiaria decumbens* under grazing 5-species Forage Bank	
Mixed Forage Bank	Forest	Maize-beans-sweet-potatoes		Fruit trees + maize	Star grass	*Trichanthera gigantea*, *Morus* spp., *Erthrina edulis*, *Boehmeria nivea*, *Tithonia diversifolia*		
Native Forest	Forest		Forest (intervened)				Forest (nonintervened)	

Area 2: Dagua (1350 m.a.s.l.), Colombia

Present Land Use System	Initial	1950's	1960 - 1986	1986 - 2002
Degraded Pasture	Forest	*Hyparrhenia ruffa* pasture under grazing		Degraded *Hyparrhenia rufa* pasture
Improved Pasture	Forest	Coffee	*H. rufa* pasture	*Brachiaria decumbens* under rotational grazing
Mixed Forage Bank	Forest	Coffee	*H. rufa* pasture	4-species Forage Bank *Trichanthera gigantea*, *Morus* spp., *Erthrina fusca*, *Tithonia diversifolia*
Native Forest	Forest	Forest (intervened)		Forest (regenerated)(naturally or anthropogenically?)

pose of this research was to identify pasture and silvo-pastoral systems that represent viable economic alternatives to the farmer apart from providing environmental services, particularly carbon sequestration. Research is conducted in three phases: (a) characterization of improved-systems farms versus conventional-systems farms in terms of environmental and socio-economic indices; (b) simulation of investment scenarios to perform cost/benefit analysis and understand the relevance for farmers to invest in carbon sequestration; and (c) formulation of policy recommendations at local, national and international levels. The first phase, the characterization of improved-systems farms versus conventional-systems farms in their environmental conditions and farmers' welfare, has concluded. Environmental, life quality and socio-economic indices per farm were produced and statistically compared between the two groups of farms. Partial results are reported in the present article. A follow-up with the participating farmers to ensure that they continue to manage these systems adequately is not the subject of the present on-going research project. However, it is expected to be the central objective of a new, future research proposal.

RESULTS

Soil Carbon Stocks

Tables 4 and 5 show statistical comparisons of SCS between land-use systems using fixed soil mass and fixed soil depth estimates in the tropical Andean hillsides of Colombia (Table 4) and in the sub-humid and humid tropical forest of Costa Rica (Table 5). When SCS estimates were adjusted to fixed soil mass, corresponding rankings between land use systems remained the same, but SCS estimates and statistical comparisons between systems changed, indicating an overestimate of stocks when using fixed-depth calculations.

For Zone 1 (Dovio), with higher altitude, steeper slopes and higher relative fertility, native forest (234 Mg/ha/1m-equivalent) had statistically higher stocks than improved *B. decumbens* pasture, degraded pasture, and mixed-forage bank for cut and carrying (162, 156 and 138 Mg/ha/1m-equivalent, respectively) (Table 4). On Zone 2 (Dagua), with lower altitude, less inclined slopes and lower fertility, lower levels of SCS were found for all systems. Native forest (186 Mg/ha/1m-equivalent) had statistically higher stocks than secondary forest, natural regeneration of a degraded pasture and improved *B. decumbens* pasture (152,

TABLE 4. Soil Carbon Stocks (Mg C/ha) per Land Use System, Estimated Based on Fixed Soil Mass (Method 1) and Fixed Soil Depth (Method 2). Tropical Andean Hillsides, Colombia.

	Zone 1: Dovio (1900 m.a.s.l.)					
System	**Total C (Mg/ha)**		**Oxid C (Mg/ha)**		**Stable C (Mg/ha)**	
(N = 12 sampling points per land use system)	Meth 1	Meth 2	Meth 1	Meth 2	Meth 1	Meth 2
1. Native forest	234 a	262 a	169 a	184 a	67 a	79 a
2. *B. decumbens* pasture	162 b	213 b	125 b	159 b	38 b	55 ab
3. Degraded pasture	156 b	183 bc	121 b	139 c	37 b	46 ab
4. Mixed forage bank	138 b	161 c	94 c	106 d	47 b	58 b
Mean	173	209	127	151	47	58
CV (%)	22.2	21.2	15.4	14.6	48.5	47.3
	Zone 2: Dagua (1350 m.a.s.l.)					
System	**Total C (Mg/ha)**		**Oxid C (Mg/ha)**		**Stable C (Mg/ha)**	
(N = 12 sampling points per land use system)	Meth 1	Meth 2	Meth 1	Meth 2	Meth 1	Meth 2
1. Native forest	186 a	214 a	149 a	172 a	42 ab	50 b
2. Secondary forest	152 b	177 b	115 b	129 b	37 b	48 b
3. Nat. regen. of degr. pasture (Fallow land)	147 b	171 b	89 c	93 c	59 a	78 a
4. *B. decumbens* pasture	142 b	165 b	118 b	141 b	35 b	42 b
5. Degraded land	97 c	125 c	62 d	71 d	35 b	54 b
6. Mixed forage bank	86 c	104 d	60 d	68 d	26 b	36 c
Mean	135	159	99	112	39	51
CV (%)	27.9	27.9	15.6	13.7	56.4	55.4

Adapted from Amézquita et al. (2003)

147 and 142 Mg/ha/1m-equivalent, respectively, not statistically different), which in turn had statistically higher stocks than degraded soil and mixed forage bank for cut and carrying (97 and 86 Mg/ha/1m-equivalent, respectively). Although native forest possesses the highest soil carbon accumulation capacity in this ecosystem, improved pasture systems and natural regeneration systems (fallow land and secondary forest) seem good environmental solutions for the recovery of degraded areas, as C-improved systems.

TABLE 5. Soil Carbon Stocks (Mg C/ha) per Land Use System, Estimated Based on Fixed Soil Mass (Method 1) and Fixed Soil Depth (Method 2). Humid and Sub-Humid Tropical Forest, Costa Rica.

Zone 1: Pocora, Atlantic Coast, Humid Tropical Forest						
System	**Total C (Mg/ha)**		**Oxid C (Mg/ha)**		**Stable C (Mg/ha)**	
(N = 12 sampling points per land use system)	Meth 1	Meth 2	Meth 1	Meth 2	Meth 1	Meth 2
1. *I. ciliare* pasture	208 a	212 a	182 a	186 a	26 a	27 a
2. *B. brizantha* + *A. pintoi*	194 a	202 a	166 b	172 a	28 a	27 a
3. *A. mangium* + *A. pintoi*	168 b	173 b	135 c	139 a	33 a	34 a
4. *B. brizantha*	134 c	135 c	120 cd	120 c	15 b	15 b
5. Native forest	128 c	128 c	111 d	111 c	17 b	17 b
6. Degraded pasture	94 d	101 d	84 e	90 d	10 b	11 b
Mean	159	158	135	136	22	22
CV (%)	27.0	26.8	17.0	18.1	49.4	48.7
Zone 2: Esparza, Pacific Coast, Sub-Humid Tropical Forest						
System	**Total C (Mg/ha)**		**Oxid C (Mg/ha)**		**Stable C (Mg/ha)**	
(N = 12 sampling points per land use system)	Meth 1	Meth 2	Meth 1	Meth 2	Meth 1	Meth 2
1. Native forest	185 a	194 a	171 a	180 a	14 a	15 ab
2. *H. rufa* pasture	169 a	180 a	153 a	162 ab	16 a	18 ab
3. *B. decumbens* pasture	134 a	137 a	104 a	106 ab	30 a	31 a
4. Forage bank	130 a	133 a	117 a	117 ab	14 a	14 ab
5. Silvo-pastoral system	130 a	132 a	112 a	119 ab	18 a	14 ab
6. Degraded pasture	129 a	195 a	117 a	126 ab	11 a	11 b
7. Secondary forest	116 a	116 a	101 a	101 b	15 a	15 ab
Mean	143	155	128	131	16	17
CV (%)	25.2	24.3	23.3	21.2	45.4	42.9

Adapted from Llanderal and Ibrahim (2004).

For Zone 1 (Pocora), a humid tropical forest on the Atlantic coast with a humid environment year round, pasture systems such as *I. ciliare*, *B. brizantha* + *A. pintoi*, *A. mangium* + *A. pintoi* and *B. brizantha* in monoculture (208, 194, 168 and 134 Mg/ha/1m-equivalent, respectively) had statistically higher stocks than native forest (128 Mg/ha/1m-equivalent), which in turn had statistically higher stocks than degraded pasture (94 Mg/ha/1m-equivalent) (Table 5). Similar rankings were obtained in the humid tropical forest of Amazonia, Colombia, where *B. humidicola* and *B. decumbens* pastures (monoculture and legume-associated) showed higher SCS than native forest (data not shown in the present paper). In Zone 2 (Esparza)–sub-humid tropical forest on the Pacific coast, with six months of severe dry season–no statistical differences were found between land-use systems in their SCS level expressed either as total carbon, oxidizable carbon or stable carbon. In hot and humid environments, improved pasture systems show SCS comparable or higher than native forest, therefore representing attractive environmental solutions for the recovery of degraded areas as C-improved systems.

Results of Multivariate Analysis

Principal Component Analysis and Cluster Analysis were used to identify relationships between SCS and soil parameters and to group sampling points with similar soil conditions and SCS level. Table 6 shows results of the tropical Andean hillsides ecosystem. Principal Component Analysis allowed for the reduction of soil and carbon parameters to two main principal components. They explained 39% and

TABLE 6. Association Between Soil and Carbon Variables. Principal Component Analysis.

Variable	PC_1 (39%)	PC_2 (28%)
	Principal Components' Scores	
Total C (Mg/ha/1m-equivalent)	0.47	0.43
Total N (Mg/ha/1m-equivalent)	0.46	0.31
Stable C (Mg/ha/1m-equivalent)	0.15	0.56
Sand (Mg/ha/1m-equivalent)	0.49	−0.32
Clay (Mg/ha/1m-equivalent)	−0.50	0.32
pH (mean in 1m-equivalent)	0.12	0.12
CEC(meq) (mean in 1m-equivalent)	0.24	−0.38

28%, respectively–67% combined–of the total variance present in the original variables.

Principal Component 1 (39% of the total variance) suggests a positive association of total carbon and total N with predominance of sand over clay. Principal Component 2 (28% of the total variance) suggests a positive association of stable carbon with predominance of clay over sand.

Cluster Analysis, using as classification criteria soil and carbon variables, allowed for the grouping of sampling points with similar soil conditions and soil carbon stock levels. Six groups were obtained explaining 77% of the total variability among sampling points. Characterization of cluster groups presented in Table 7, suggests that although land-use system and site conditions seem to be the most important factors determining SCS, other factors not considered in this study play an important role in soil carbon accumulation, such as prior land-use history of each field.

TABLE 7. Soil Carbon Stocks by Cluster Using 93 Sampling Points, Grouped in 6 Clusters.

Cluster	Total C (Mg/ha/1m-equiv) Min-Max (Mean)	Stable C (Mg/ha/1m-equiv) Min-Max (Mean)	Sampling Points in the Cluster
1 (N = 5)	300-374 (335)	66-131 (86)	Forest Zone 1 (3) Forest Zone 2 (2)
2 (N = 8)	203-287 (248)	23-108 (68)	Forest Zone 1 (3) Forest Zone 2 (3) Improved pasture Zone 1 (1) Improved pasture Zone 2 (1)
3 (N = 30)	152-299 (211)	23-129 (68)	All points belong to Zone 1 Forest (5), Impr. past. (11), Degr. past (9), Forage Bank (8)
4 (N = 21)	118-239 (171)	23-148 (72)	All points belong to Zone 2 Improved pasture (9) Degraded pasture (12)
5 (N = 6)	171-192 (160)	8-52 (24)	Forest Zone 1 (1) Forest Zone 2 (5)
6 (N = 23)	70-171 (124)	7-71 (34)	Forage Bank Zones 1, 2 (13) Degr. past. Zone 1 (6) Impr. past. Zone 2 (2) Forest Zone 2 (2)

TABLE 8. Socio-Economic Indices of Improved vs. Conventional Farms. Tropical Andean Hillsides, Colombia.

Index	Farm Type		
	Improved (N = 6)	Conventional (N = 19)	P
1. Farm area in forest (%)	29	14	**
2. Farm area improved systems (%)	88	44	**
3. Farm gross income/ha/yr (US $)	250	50	***
4. Farmer self-sufficiency (%)	40	32	*
5. Living conditions (1-5)	5	3	**
6. Educational level			
• Adult literacy (%)	79	76	*
• Mean years of schooling	8	6	*

P = Probability of statistical significance; * : 0.05 < P < 0.10; ** : 0.01 < P < 0.05; *** : P < 0.01

Socio-Economic Indicators

In order to compare improved versus conventional farms in their provision of environmental services and socio-economic benefit to the farmer, environmental and socio-economic indicators were estimated per farm for the four ecosystems. Table 8 shows that for the tropical Andean hillsides of Colombia, improved-systems farms perform statistically better than conventional-systems farms, both in environmental conditions, such as percent area forested and percent area in improved systems, and in socio-economic conditions, such as farm gross income/ha/year, farmer self-sufficiency, family living conditions, and educational level.

CONCLUSIONS

- In the tropical ecosystems of Latin America studied in the present research, improved pasture and silvo-pastoral systems show SCS levels comparable or even higher than those from native forest, depending on climatic and environmental conditions (altitude, temperature, precipitation, topography and soil). Therefore, these systems should be considered as attractive and viable C-improved systems.

- Carbon sequestration research requires the use of an appropriate methodology for field evaluation and mathematical estimation of soil carbon stocks. The following factors should be taken into account (a) a soil sampling design taking into account factors affecting SCS needs to be used to obtain minimum-variance estimates; (b) variability of SCS estimates depends on land-use type (i.e., higher variability on degraded pasture systems, characterized by a high heterogeneity in vegetation, than on improved grass-alone systems or silvo-pastoral systems with less heterogeneity in vegetation), site conditions (altitude, climate, topography), soil characteristics, and carbon fraction; (c) for statistical comparisons of land-use systems, SCS estimates corrected by bulk density and adjusted to fixed soil mass per sampling point should be used, which are more precise than those based on fixed soil depth calculations; and (d) SCS estimates for a given land-use system need to be interpreted based on long-term land -use history.
- Tropical pasture and silvo-pastoral systems are important socio-economic components of the economies of Tropical America's countries, across all ecosystems. When improved and well managed, they can become key land-use systems for the provision of environmental services, particularly the recovery of degraded areas and carbon sequestration. In addition, they have the capacity to provide viable economic alternatives to farmers.

REFERENCES

Amézquita, E., J. Ashby, E.K. Knapp, R. Thomas, K. Muller-Samann, H. Ravnborg, J. Beltran, J.I. Sanz, I.M. Rao, and E. Barrios. 1998. CIAT's strategic research for sustainable land management on the steep hillsides of Latin America. In F.W.T Penning de Vries, F. Agus and J. Kerr (eds). "Soil Erosion at Multiple Scales," CAB INTERNATIONAL: Pp. 121-131.

Amézquita, M.C. 2003. Evaluation and Análisis of Carbon Stocks in Pasture, Agro-pastoral and Silvo-pastoral Systems in Sub-ecosystems of the American Tropical Forest. In M.C. Amézquita and F. Ruiz (eds). Two-year Project Achievements. Internal Publication No. 9, Carbon Sequestration Project The Netherlands Cooperation CO-010402, Fourth International Coordination Meeting. December 2003, CIAT, Cali, Colombia.

Amézquita, M.C. 2002. Project objectives, expected products and research methodology. In M.C. Amézquita, F. Ruiz and B. van Putten (eds). Carbon Sequestration and Farm Income: Concepts and Methodology. Internal Publication No. 5, Carbon Sequestration Project The Netherlands Cooperation CO-010402, Second International Coordination Meeting, December 2002, CATIE, Turrialba, Costa Rica.

Amézquita, M.C., M. Ibrahim and P. Buurman. 2004. Carbon sequestration in pasture, agro-pastoral and silvo-pastoral systems in the American Tropical Forest Ecosystem. In Proc. 2nd Intl. Congress in Agroforestry Systems, Mérida, Mexico, February 2004: 61-72.

Amézquita, M.C., H.F. Ramírez, E. Amézquita, H. Giraldo and M.E. Gómez. 2003. Carbon storage in long-established systems: Two-year research results, Andean Hillsides, Colombia. In M.C. Amézquita, and F. Ruiz (eds) Two-year Project Achievements. Project Internal Publication No. 9, Carbon Sequestration Project The Netherlands Cooperation CO-010402, Fourth International Coordination Meeting. December 2003, CIAT, Cali, Colombia.

Browder, J.O. 1988. The social costs of rain forest destruction: A critique and economic analysis of the "hamburger debate." Interciencia 13(3):115-120.

Buurman, P., M. Ibrahim and M.C. Amézquita. 2004. Mitigation of greenhouse gas emissions by silvopastoral systems: Optimism and facts. In Proc. 2nd Intl. Congress in Agroforestry Systems, Mérida, Mexico, February 2004.

CIAT (Centro Internacional de Agricultura Tropical). 1976-2000. Tropical Pastures Program and Tropical Forages Project Annual Reports. Cali, Colombia.

Ellert, B.H., H.H. Janzen and T. Entz. 2002. Assessment of a method to measure temporal change in soil carbon storage. Soil Sci. Soc. Am. J. 66:1687-1695.

FAO. 2000. Food balance sheets. Rome, Italy.

Jones, P. 1993. Hillsides Definition and Classification. CIAT, Cali, Colombia, 7 pp.

Kaimowitz, D. 1996. Livestock and deforestation Central America in the 1980s and 1990s: A policy perspective. Center for International Forestry Research (CIFOR), Special Publication, Jakarta. pp. 88.

Llanderal, T. and M. Ibrahim. 2004. Biophysical Analysis: Advancement Report Sub-humid and humid Tropical Forest, Costa Rica. In Six-months Report No. 5, Internal Document No. 11, Carbon Sequestration Project, The Netherlands Cooperation CO-010402, Cali, Colombia.

Pachico, D., J. Ashby and L.R. Sanint. 1994. Natural resource and agricultural prospects for hillsides of Latin America. Paper prepared for IFPRI 2020-Vision. Workshop Washington, 7-10 November. In Hillsides Program Annual Report 1993-1994, pp. 283-321. CIAT, Cali, Colombia.

Ramírez, C. and R. Ortiz. 1997. Causas de pérdida de biodiversidad. In "Informe Nacional sobre el estado de la biodiversidad Colombia," Tomo II, pp. 33-35. Instituto de Investigación de recursos biológicos Alexander von Humboldt.

Rivas, L., D. Pachico, C. Seré and J. García. 1998. Evolución y perspectivas de la ganadería vacuna en América Latina Tropical en un contexto mundial. Proyecto de Evaluación de Impacto. CIAT. Cali, Colombia.

Rubinstein, E. de and G.A. Nores. 1980. Gasto en carne de res y productos lácteos por estrato de ingreso en doce ciudades de América Latina. CIAT. Cali, Colombia.

Sader, S.A. and A.T. Joyce. 1988. Deforestation and trends in Costa Rica, 1940 to 1983. Biotropica 20(1): 11-19.

Sanint, L.R., L. Rivas, M.C. Duque and C. Seré. 1984. Food consumption patterns in Colombia: A cross sectional analysis 1981. Paper presented at the Internal Workshop of Agricultural Centers on Selected Economic Research Issues in Latin America. CIAT. Cali, Colombia.

Toledo, J.M. 1985. Pasture development for cattle production in the major ecosystems of the tropical American lowlands. In Proc. of the XV Intl. Grasslands Congress, pp. 74-81. Kyoto, Japan.

USDA. 1996. Soil survey laboratory methods manual. Soil Survey Investigations Report No. 42, Version 3, United States Department of Agriculture, Washington, DC, USA, 693 pp.

Valdés, A. and G.A. Nores. 1979. Growth potential of the beef sector in Latin America–survey of issues and policies. In 4th World Conference of Animal Production. Buenos Aires, Argentina.

Veldkamp, E. 1993. Soil organic carbon dynamics in pastures established after deforestation in the humid tropics of Costa Rica. PhD Thesis, pp. 117. Wageningen Agricultural University. Wageningen, The Netherlands.

Veldkamp, E. 1994. Organic carbon turnover in three tropical soils under pasture after deforestation. Soil Sci. Soc. Am. J. 58: 175-180.

Vera, R., J.I. Sanz, P. Hoyos, D.L. Molina, M. Rivera and M.C. Moya. 1993. Pasture establishment and recuperation with undersown rice on the acid soil savannas of South America. CIAT. Cali, Colombia.

Environmental Services of Native Tree Plantations and Agroforestry Systems in Central America

Florencia Montagnini
Daniela Cusack
Bryan Petit
Markku Kanninen

SUMMARY. Besides supplying the growing demand for wood, plantations and agroforestry systems provide environmental services such as carbon sequestration and recovery of biodiversity. Several countries of Central America have recently started incentive programs to encourage plantation and agroforestry development. In Costa Rica, Payment for Environmental Services (PES) provides subsidies to farmers for plantations and agroforestry systems. Funding for these subsidies comes

---

Florencia Montagnini is Professor in the Practice of Tropical Forestry, Yale University, School of Forestry and Environmental Studies, 370 Prospect Street, New Haven, CT 06511 USA (E-mail: florencia.montagnini@yale.edu).

Daniela Cusack is Doctoral Student, University of California-Berkeley, Environmental Science, Policy and Management, 151 Hilgard Hall, Berkeley, CA 94720 USA (E-mail: dcusack@nature.berkeley.edu).

Bryan Petit is Project Leader, San Diego County/Forester, USDA, Natural Resources, Conservation Service, Southern California Watershed Recovery Program, 332 South Juniper Street, NRCS, #110, Escondido, CA 92026 USA (E-mail:.bryan.petit@ca.usda.gov).

Markku Kanninen is affiliated with the Center for International Forestry Research (CIFOR), Jalan CIFOR, Situ Gede, Sindangbarang, Bogor Barat 16680, Indonesia (E-mail: m.kanninen@cgiar.org).

[Haworth co-indexing entry note]: "Environmental Services of Native Tree Plantations and Agroforestry Systems in Central America." Montagnini, Florencia et al. Co-published simultaneously in *Journal of Sustainable Forestry* (Food Products Press, an imprint of The Haworth Press, Inc.) Vol. 21, No. 1, 2005, pp. 51-67; and: *Environmental Services of Agroforestry Systems* (ed: Florencia Montagnini) Food Products Press, an imprint of The Haworth Press, Inc., 2005, pp. 51-67. Single or multiple copies of this article are available for a fee from The Haworth Document Delivery Service [1-800- HAWORTH, 9:00 a.m. - 5:00 p.m. (EST). E-mail address: docdelivery@haworthpress.com].

Available online at http://www.haworthpress.com/web/JSF

doi:10.1300/J091v21n01_03

from a special tax on gasoline, and from external sources sought by FONAFIFO (National Fund for Forestry Financing).

These plantations and agroforestry systems are established on degraded land by farmers who are often advised by local NGOs (non-governmental organizations) or by research institutions that have practical experience in the region. Gathering information on species selection, plantation silviculture, and environmental services provided by plantations and agroforestry systems is important to the success of these systems. These incentive programs can also serve as a model for starting or modifying similar programs in other countries with comparable ecological and socioeconomic conditions.

This paper presents experiences with native species plantations over the past twelve years in Costa Rica. Data on volume, biomass production and recuperation of biodiversity are presented. We recommend the establishment of government incentives for reforestation and agroforestry systems with native species.

KEYWORDS. Biodiversity, carbon sequestration, degraded lands, mixed plantations, native species, reforestation, subsidies

INTRODUCTION

Tropical plantations provide a variety of environmental services such as wood production, carbon sequestration, and acceleration of forest succession processes (Parrotta, 1992; Lamb, 1998). With relatively high yields, tropical and subtropical plantations have the potential to contribute substantially to global wood production (Evans, 1999; Wadsworth, 1997).

In Central America, results of experiments started in the 1980s have helped to identify promising native and exotic species for reforestation and agroforestry. For example, the Proyecto Madeleña (Timber and Fuelwood Project) of CATIE (Centro Agronómico Tropical de Investigación y Enseñanza) (Tropical Agriculture Research and Higher Education Center) in Costa Rica, focused on six countries: Panama, Guatemala, Nicaragua, Costa Rica, El Salvador and Honduras (Ugalde, 1997). CATIE and other institutions like the Organization for Tropical Studies (OTS), and the Instituto Tecnológico de Costa Rica (Costa Rican Technological Institute) (ITCR) have generated valuable information about growth, productivity, biomass, and financial aspects for native and exotic trees in pure and mixed-species plantations of the humid and sub-humid tropical regions of Central America. The species studied have been estimated to have rotation periods of 12-25 years and projected yields of 250-300 m^3/ha (González and Fisher, 1994; Butterfield and Espinoza, 1995; Montagnini et al., 1995; Montagnini and Mendelsohn, 1997; Haggar et al., 1998; Petit and Montagnini, 2004).

Planting the studied tree species is an attractive alternative for farmers. Firewood from thinnings is an additional source of income. Recently, interest in establishing mixed plantations and agroforestry systems including native tree species has grown among small farmers in Central America (Montagnini et al., 1995; Montagnini and Mendelsohn, 1997; Piotto et al., 2003a; 2003b). Farmers have interest in planting native species because many of the native species whose timber has wide acceptance and an established demand in local markets are becoming increasingly scarce in natural forests due to logging, or in some cases because their extraction has been banned recently. They consider reforesting their farms with valuable species as a viable alternative to past land uses, such as extensive cattle ranching, which became unprofitable and degraded their lands. Most farmers who own small and medium-sized farms are often willing to allocate limited portions of their land to native species, while keeping major portions of their properties for other land uses, including reforestation with better known exotic species, such as *Tectona grandis* (teak) and *Gmelina arborea* (Piotto et al., 2003a; Piotto et al., 2004).

In Costa Rica, the forestry legislation includes incentives for the establishment and management of plantations and agroforestry systems, especially on abandoned pastures and other deforested lands. Due to these incentives, interest in establishing plantations and agroforestry systems has grown among farmers. Nonetheless, farmers and entrepreneurs need better technical information about the appropriate silviculture practices for native species to increase the commercial value of these production systems. More information would be especially valuable for improved plantation and agroforestry systems planning and management.

This paper presents results from experimental native tree plantations in Costa Rica, where growth in mixed and pure-species designs, carbon sequestration, and recuperation of biodiversity were measured. Government incentives for native-species plantations and agroforestry systems are recommended for countries with ecological and socioeconomic conditions similar to those in Costa Rica.

DEVELOPMENT OF PLANTATIONS IN CENTRAL AMERICA

The total area of plantations in Central American countries is shown in Table 1. Costa Rica and Guatemala are the two Central American countries with the largest area of plantations; they also have the oldest government programs that subsidize reforestation, which were started in the early 1980s. These programs generally work in association with non-government organizations (NGOs) that provide technical assistance to farmers in both silvicultural aspects and procedures for obtaining subsidies.

Historically, the wood market in Costa Rica has relied on the exploitation of natural forests. Commercially, wood primarily has been used for furniture, doors for export, and house construction. Reforestation using native species has been evolving over the past two decades. In Costa Rica, various government incentives exist for the establishment and management of plantations and agroforestry systems. Incentives include Payment for Environmental Services (PES) and advanced purchase of timber. With Forestry Law #7575, Costa Rica officially recognized certain environmental services provided by natural forests and plantations in 1996. The environmental services recognized in Costa Rica are defined in the Forestry law #7575: (a) mitigation of emissions of greenhouse gases (carbon fixation and storage); (b) protection of water for urban, rural or hydroelectric use; (c) protection of biodiversity

TABLE 1. Total area in tree plantations in Central America.

Country	Plantation area (hectares)	Size of the country (km^2)
Costa Rica	178,000	51,060
Guatemala	133,000	108,430
Honduras	48,000	111,890
Nicaragua	46,000	74,430
El Salvador	14,000	20,720

Source: FAO (2001).

for conservation and sustainable use for scientific, medicinal, research, or genetic uses, by protecting ecosystems and life forms; and (d) natural scenic beauty for tourist and aesthetic purposes. Methods of payment for these services were also established (Campos and Ortíz, 1999). This law allows for payments to owners of forested land and/or plantations in recognition of the environmental services provided. The government through selective gasoline taxes primarily finances the incentives programs, although international sales of carbon credits and loans also contribute to finance the program.

By providing monetary incentives to private land owners, PES could encourage numerous and extensive reforestation projects. PES are made to owners of 2-300 hectares for the conservation and regeneration of natural forest. Owners of more than one hectare are paid for reforestation or plantation establishment. Applications are made at regional offices through the National Fund for Forest Finance (FONAFIFO), which manages funds and issues certificates for PES. Payments are dispersed over five years at varying levels, depending on the type of forestry activity (Campos and Ortíz, 1999). PES are highest for reforestation (establishing of new plantations, generally on abandoned lands from agriculture or cattle), with payments totaling over $600 per hectare over a five-year period, which is almost twice the amount paid for conservation of natural forests. PES for agroforestry systems are given at a rate of $1.00 per tree. PES for plantation establishment could be complemented with other avenues of income generation, such as the advanced purchase of timber. Advanced purchase of timber is generally done by FUNDECOR (Foundation for Development of the Central Mountain Range), a NGO that assists farmers in the technical aspects of forest management and reforestation in the northern region of Costa Rica. This organization pays landowners in advance an average of $75 ha^{-1} yr^{-1} for up to fifteen years for plantations over three-years old. In addition to the annual payments, the landowners receive 80% of the profit from the final timber sale.

Agricultural or cattle ranching activities can generate income soon after the establishment of agroforestry systems, accelerating the recovery of costs incurred by the initial investment. In addition, PES and advanced purchase of timber ameliorate to varying extents one of the primary bottlenecks of reforestation: the high cost of plantation establishment and management.

Selection of Species for Reforestation of Degraded Land

Plantations and agroforestry systems can provide forest products (wood, firewood, mulch) and ecological benefits, such as improved nutrient cycling, soil conservation, and recovery of biodiversity. Species selection depends on the possibility of obtaining the desired products as well as achieving environmental benefits.

In Costa Rica, research on the growth of native and exotic tree species has been conducted since the early 1980s by a number of institutions such as CATIE, OTS, ITCR, other universities, and several international or local development projects. As a result of this research, scientists have been able to indicate which tree species are best for reforestation and agroforestry systems. For example, in the Atlantic humid lowlands of Costa Rica recent experiments have tested growth and agroforestry potential of several native tree species, such as *Terminalia amazonia*, *Virola koschnyi*, *Dipteryx panamensis*, *Vochysia ferruginea*, *Vochysia guatemalensis*, *Hieronyma alchorneoides*, and *Calophyllum brasiliense* (Montagnini et al., 1995; Montagnini and Porras, 1998; Byard et al., 1996; Kershnar and Montagnini, 1998; Horn and Montagnini, 1999; Piotto et al., 2003a; 2003b; Petit and Montagnini, 2004). All these species are frequently grown by farmers as part of plantations or agroforestry systems subsidized by the PES program (Piotto et al., 2003a).

GROWTH AND PRODUCTIVITY OF MIXED AND PURE-SPECIES PLANTATIONS

In Central America, many plantations and agroforestry systems are being established on abandoned agricultural land. Farmers are recognizing the potential for financial returns from plantations. However, these farmers prefer to use native species, of which relatively little is known (Piotto et al., 2003b). While establishment considerations and management practices are known for some native species, productivity levels of native species in plantations and agroforestry systems are mostly unknown (Montagnini et al., 1995; Piotto et al., 2003a). The next step in finding a suitable species for reforestation efforts is to determine the length of each species' rotation cycle and its timber production capacity.

As researchers continue to demonstrate the benefits of native species and mixed-species plantations, they will also continue to influence the method and type of plantations and agroforestry systems being implemented by the farmers. The diversity of species in plantations (both native and exotic species) has been increasing (FAO, 2001). Also, many local farmers seek to use mixed-species plantations (Piotto et al., 2003b). Well-planned, mixed-species plantations provide more diverse products than pure-species plantations. Mixed-species plantations diminish market risk, reduce incidence and severity of certain pathogen attacks, and complement ecosystem resource use (Wormald. 1992; Montagnini et al., 1995).

Volume Yield and Rotation Times of Native Species at La Selva, Costa Rica

For twelve years, we measured growth and productivity of twelve native species on three experimental plantations growing in mixed and pure-species plots at La Selva Biological Station in the humid Atlantic lowlands of Costa Rica. The twelve native species studied were: Plantation 1: *Jacaranda copaia* (Aubl.) D. Don, *Vochysia guatemalensis* D. Sm., *Calophyllum brasiliense* Cambess, and *Stryphnodendron microstachyum* Poepp. et Endl.; Plantation 2: *Terminalia amazonia* (Gmell.) Exell., *Dipteryx panamensis* (Pittier) Record & Mell, *Virola koschnyi* Warb, and *Paraserianthes guachapele* (Kunth) Harms; Plantation 3: *Hieronyma alchorneoides* Fr. Allemao, *Balizia elegans* (Ducke) Barnaby and Grimes, *Genipa americana* L., and *Vochysia ferruginea* Mart. The plantations were designed based on species growth, form, economic value, farmer preference, and potential impact on soil fertility recuperation (Montagnini et al., 1995). Plantations were in completely randomized blocks with four replications and five treatments: four pure-species plots per species and one mixed-species plot with all four species. Initial spacing was at 2 meters × 2 meters, and each plot was 32 meters × 32 meters with a total of 256 trees per plot. Thinnings at years 3 and 6 left trees at 4 meters × 4 meters spacing. In each mixed-species plot, a systematic design was used to maximize species interactions (Montagnini et al., 1995). In each plot, individuals from the four species were alternated by row.

Measurements performed in 2002 were done on ten surviving species (all of those mentioned above except for *S. microstachyum* and *P. guachapele*, which had suffered severe pest attacks). Growth equations were constructed for each of the ten species. The growth equations of six of the species were third degree equations and, thus, could be used for volume extrapolations. These six species included *C. brasiliense*, *V.*

guatemalensis, *J. copaia*, *V. ferruginea*, *G. americana*, and *B. elegans*. Rotation ages and expected merchantable volumes were calculated from the growth equations (Table 2).

Yield projections are still being refined because thinning compensations have exaggerated predicted harvestable volumes for some species. However, trends can still be extrapolated from the results. As calculated from the results of the experiment, *J. copaia* has a relatively short rotation age of 6.5 years, after which trees of this species slow their growth quickly in a manner typical of most fast-growing species. At this age, trees of this species are of merchantable size and can be harvested. The rotation ages for the other four species range from 13.2 to 18.5 years (Petit and Montagnini, 2004).

Two of the three species comprising Plantation 1, *V. guatemalensis* and *J. copaia*, grew faster in mixtures than in pure plantations, while *C. brasiliense* grew significantly slower in the mixed-species plantations. We determined the ideal rotation age of the mixed-species plantations to be 8.5 years and its corresponding yield to be 403 m^3/ha. Subsequently, we compared the productivity of the mixed-species plantation to that of each of the single-species plantations. The mixed-species plantation produced more merchantable wood over time than any of the other pure plantations from this experiment. The mixture was approximately 21% more productive than the most productive of the single-species plantations (Petit and Montagnini, 2004).

In Plantation 2, *Virola koschnyi*, *Terminalia amazonia*, and the mixed-species plantations had the highest growth increments in volume. Plots with the highest productivity in Plantation 3 were *Vochysia ferruginea*,

TABLE 2. Rotation age and merchantable volume of six species at La Selva Biological Station, Costa Rica.

Species	**Rotation Age (years)**	**Merchantable volume at time of harvest (m^3/ha)**
C. brasiliense	18.5	296
V. guatemalensis	13.5	417
J. copaia	6.5	255
V. ferruginea	13.3	363
G. americana	13.2	76
B. elegans	8.8	139

followed by the mixed-species plantations, *Balizia elegans,* and *Hieronyma alchorneoides* (Piotto et al., 2003b; Petit and Montagnini, 2004).

Mixed plantations, if planned according to the characteristics of each species, can in general produce more wood than pure-species plantations. Mixtures of fast-growing species and slower-growing species produce harvestable wood at different rotation times, with faster growing species producing timber of less value (e.g., *Jacaranda copaia*) and slower-growing species (e.g., *C. brasiliense, D. panamensis*) producing more valuable wood. The more valuable wood also constitutes a longer-term sink for fixed carbon (e.g., construction timber, furniture, wood crafts) than timber of less value, whose uses may be relatively shorter-lived (e.g., boxes, poles, fuelwood). As different species have different rotation times, the land in a mixed-species plantation is in use for a longer period of time than if planted with just one fast-growing, short rotation species. This diminishes incentives for changing to other land uses, keeps a vegetative cover that protects the soil, and serves other environmental services such as biodiversity conservation as well (Montagnini and Porras, 1998).

Biomass and Carbon Accumulation in Mixed and Pure-Species Plantations

As determined from measurements of biomass from thinnings at ten years of age in Plantation 1, pure plantations of *Vochysia guatemalensis* had the highest biomass, followed by *Calophyllum brasiliensis* (Table 3). In Plantation 2, the highest biomass per hectare was measured on *Dipteryx panamensis* and *Terminalia amazonia*. On Plantation 3, *Hieronyma alchorneoides* had the highest biomass, followed by *Vochysia ferruginea*.

From the aforementioned results, we have determined which species are the best options for increasing biomass accumulation and carbon sequestration in plantations. The values of mean annual aboveground biomass production for the fastest-growing species of our experiments lie within the ranges reported elsewhere for fast-growing plantations of commonly-used exotics in the humid tropics (Montagnini and Porras, 1998). Values for the two slower growing trees in pure plots, *C. brasiliense* and *D. panamensis,* are similar to ranges reported for relatively slower-growing species.

This information can be used for programs aimed at compensating for carbon sequestration by specific land uses (plantations or agroforestry). We are currently using a simulation model (CO2FIX) to estimate carbon sequestration for each species and plantation mixtures (Masera et al., 2003; Nabuurs et al., 2002).

TABLE 3. Above-ground biomass (Mg/ha) of ten native tree species in plantation at La Selva Biological Station, Costa Rica.

	Foliage	Branch	Stem	Total
Plantation 1 (10 years old)				
Calophyllum brasiliense	**11.7**	**21.5**	**59.4**	**92.6**
standard deviation	*8.4*	*13.6*	*31.3*	*52.9*
Vochysia guatemalensis	**4.4**	**10.3**	**110.9**	**125.6**
standard deviation	*2.5*	*14.8*	*80.7*	*84.3*
Jacaranda copaia	**1.5**	**3.2**	**85.0**	**89.8**
standard deviation	*1.3*	*5.7*	*46.8*	*18.2*
Plantation 2 (10 years old)				
Dipteryx panamensis	**10.5**	**30.7**	**164.0**	**205.3**
standard deviation	*5.9*	*19.8*	*84.8*	*106.9*
Terminalia amazonia	**9.3**	**29.1**	**126.5**	**164.9**
standard deviation	*4.9*	*16.4*	*69.6*	*82.6*
Virola koschnyi	**3.9**	**8.2**	**61.9**	**74.1**
standard deviation	*2.3*	*5.5*	*32.2*	*39.6*
Plantation 3 (9 years old)				
Genipa americana	**2.1**	**9.6**	**35.0**	**46.6**
standard deviation	*2.1*	*12.1*	*26.7*	*40.8*
Hieronyma alchorneoides	**4.7**	**27.1**	**87.6**	**119.4**
standard deviatin	*3.2*	*18.0*	*43.2*	*62.5*
Balizia elegans	**1.3**	**5.1**	**34.6**	**41.0**
standard deviation	*1.4*	*5.9*	*19.4*	*26.3*
Vochysia ferruginea	**5.4**	**11.5**	**61.3**	**78.1**
standard deviation	*3.9*	*10.6*	*32.0*	*44.4*

CONTRIBUTION OF PLANTATIONS TO CONSERVATION AND RECUPERATION OF BIODIVERSITY

At La Selva Biological Station, several studies have reported that plantations can accelerate the recuperation of plant diversity in the understory (Guariguata et al., 1995; Powers et al.,1997; Montagnini et al., 1999; Carnevale and Montagnini, 2002; Cusack and Montagnini, 2004). In Plantation 1, at 7 years of age, *Vochysia guatemalensis,* the mixed-species plantations, and *Calophyllum brasiliense,* had the highest abundance of woody regeneration in the understory. In Plantation 2, also at seven years, *Terminalia amazonia, Virola koschnyi* and the mixed-species plantation had the highest diversity of woody species regenerating in the understory. In Plantation 3, the highest number of woody species was under the mixed-species plantation, followed by *Hieronyma alchorneoides* and *Vochysia ferruginea* (Montagnini, 2001).

These are promising indicators for the use of these plantations as accelerators of natural forest succession in the region. In our research, tree plantations facilitated forest regeneration, influencing numbers of tree individuals, as well as species diversity. There was a larger abundance of individuals and species diversity in the mixed than in the pure plantations. Birds and bats had dispersed the majority of the regenerating individuals. The different tree species in the plantations generated different shading and litter accumulation conditions, influencing germination and survival of regenerating individuals. Species selection in reforestation projects influences the proportion of regeneration in each phase (colonization, establishment, growth and survival) (Carnevale and Montagnini, 2002).

Very few woody species were found in the control, natural regeneration plots. The lack of perches for seed dispersers and invasion by herbaceous vegetation that outcompetes the tree seedlings in their growth are factors that might have impeded the establishment of woody species in the control. In another of our studies, we compared understory woody regeneration on 10-year-old, medium-sized (~15 ha) plantations belonging to farmers in the Atlantic humid lowlands of Costa Rica. Two plantations belonging to regional farmers were compared with results from La Selva Biological Station. In this study, blocks of six native timber species were present at each of the sites. In accordance with previous results, understory regeneration was significantly higher on planted plots than on abandoned (control) pastures at each site, which were dominated by ferns (Cusack and Montagnini, 2004). Surprisingly, a different timber species was most successful for recruiting understory regeneration at each site (Figure 1). At La Selva Biological Station, *Vochysia guatemalensis* had significantly higher understory woody regeneration than other species. At the first farmer's plantation, *Terminalia amazonia* was most successful, while *Vochysia ferruginea* had the most regeneration at the second farmer's plantation. Site, more

FIGURE 1. Understory woody regeneration for the most successful timber species at La Selva Biological Station and two private farms in the Atlantic humid lowlands of Costa Rica. Cb: *Calophyllum brasiliense;* Ta: *Terminalia amazonia*; Vf: *Vochysia ferruginea.*

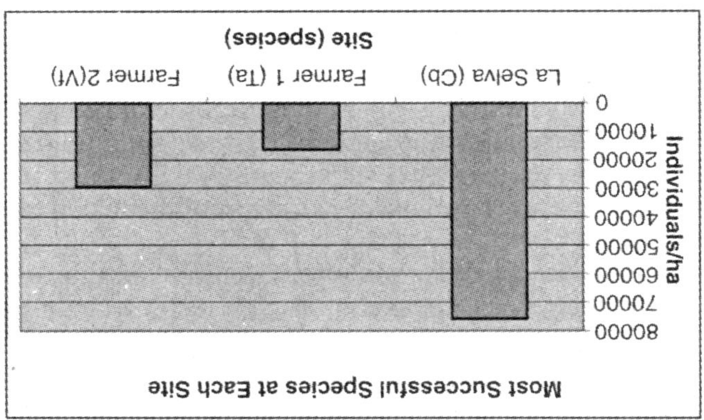

so than a particular species, can be an important factor influencing the success of planted species at recruiting understory regeneration.

In this study, we also compared dispersal vectors for regenerating species present at each site. At La Selva Biological Station, there were significantly more woody species dispersed by birds and bats than at the other two sites. Also, more individuals per hectare were present at the La Selva plantations than at the farmers' plantations (Figure 1). The high level of bird and bat dispersed species at La Selva implies that these plantations were being used as habitat by these animals. The fact that there were more bird and bat dispersed species and more individuals per hectare at La Selva Biological Station might be attributed to the proximity of these plantations to non-fragmented forest, while the farmers' plantations were farther away from natural forests.

CAN ENVIRONMENTAL SERVICES PAY FOR REFORESTATION IN CENTRAL AMERICA?

From the studies discussed here, we have found that native-species plantations have valuable social and economic functions, including the provision of forest products, carbon sequestration, and recuperation of biodiversity. Government policies that promote reforestation through incentives or other means are necessary for the successful reforestation of degraded areas. It is recommended that incentives promoting reforestation through plantation establishment and management follow similar models to those that have been successful in Costa Rica. It is also recommended that, in addition to establishment of new plantations, owners of existing plantations be allocated incentives on a yearly basis to continue proper management. The Forest Service of each country should control the payment of incentives. NGOs should assist landowners with species selection and technical aspects of plantation establishment and management.

Finding low-cost methods to sequester carbon is emerging as a major international policy goal in the context of increasing concerns about global climate change. Mitigating the accumulation of carbon dioxide and other greenhouse gases in the atmosphere through forest conservation and management was discussed as early as in the 1970s. It was not until the 1990s that international action was initiated in this direction. In 1992, several countries agreed to the United Nations Framework Convention on Climate Change (UNFCCC), whose major objectives were: developing national inventories of greenhouse gas emissions and sinks, and reducing the emission of greenhouse gases (FAO, 2001). At the third meeting of the FCCC in 1997 in Kyoto, Japan, the participating countries, including the United States, agreed to reduce greenhouse gas emissions to 5% or more below 1990 levels by 2012 (*http://unfcc.int*). The Kyoto Protocol, which emerged from this meeting, provides a mechanism by which a country that emits carbon in excess of agreed-upon limits can purchase carbon offsets from a country or region that manages carbon sinks. Although the United States' withdrawal from the treaty in 2001 has considerably weakened its implementation, the Kyoto Protocol represents a major international effort related to carbon sequestration (Montagnini and Nair, 2004). The adoption of the Kyoto Protocol in 1997 triggered a strong increase in investment in plantations as carbon sinks, although the legal and policy instruments and guidelines for management are still debated (FAO, 2000). A number of countries have already prepared themselves for the additional funding for the establishment of human-made forests. The 1997 Costa Rica national program was the first to establish tradeable securities of carbon sinks that could be used to offset emissions and to utilize independent certification insurance.

Can environmental services pay for reforestation in Central America? In Costa Rica, US $14 million were invested in the payment for environmental services in 1997, which resulted in the reforestation of 6,500 ha, the sustainable management of 10,000 ha of natural forests, and the preservation of 79,000 ha of private natural forests (Nasi et al., 2002). Eighty percent of this funding originated nationally from the tax on fossil fuels, while the other 20% came from the international sale of carbon from public protected areas. Recently, the World Bank provided a US $ 32.6 million loan to Costa Rica to fund the PES through a project called 'Ecomarkets' and was accompanied by a grant from the Global Environment Facility (GEF) of approximately US $ 8 million (Nasi et al., 2002).

In Costa Rica, the demand for PES is much higher than the funding currently available; it is estimated that the funding available could only cover 15-30% of the demand (Nasi et al., 2002). This suggests that forest owners are willing to take payments for the environmental services provided by their forests, even if probably these payments do not always fully compensate the farmers for their investment. According to recent studies, most of Costa Rican society is willing to internalize the costs of maintaining the ecological functions and environmental services from forest ecosystems, especially in the case of water. In a study carried out by CATIE, it was found that most Costa Ricans agree to pay in the form of taxes for the environmental services provided by forests. It was also found that the environmental services valued most by Costa Ricans is water protection, followed by biodiversity protection, mitigation of greenhouse gases, and scenic beauty (35%, 25%, 20% and 20%, respectively) (Nasi et al., 2002).

Apparently, the public is willing to even make additional payments for water consumption, as is currently done by the private water company, Servicios Públicos de Heredia (Heredia Public Water Company). This company charges the consumer an additional tax based on water consumption. These funds are used by the company to pay farmers who own land where the water company operates for forest protection and reforestation. The company also uses the funds to purchase affordable land that holds water reservoirs.

CONCLUSIONS

Mixed-species plantations, if planned according to species characteristics, can produce higher wood volume than pure-species plantations. Fast and slow-growing species in the same plantations have rotations of varying lengths, and they yield short and long-term products.

In the plantations studied, tree regeneration in the understory was more successful in plantations than in abandoned pastures. Mixed-species plantations demonstrated good results for recuperating understory biodiversity. As seen in these studies, native-species plantations have social and economic functions, provide forest products, contribute to the rehabilitation of degraded areas, promote atmospheric carbon sequestration, and restore biodiversity.

Government incentives or other methods are needed to promote reforestation of degraded lands. It is recommended that incentives for the establishment and management of native-species plantations follow the successful Costa Rica model.

REFERENCES

Butterfield, R. P. and M. Espinoza. 1995. Screening trial of 14 tropical hardwoods with an emphasis on species native to Costa Rica: Fourth year results. *New Forests* 9: 135-145.

Byard, R., K. C. Lewis and F. Montagnini. 1996. Leaf litter decomposition and mulch performance from mixed and monospecific plantations of native tree species in Costa Rica. *Agriculture, Ecosystems and Environment* 58:145-155.

Campos, J. J. and R. Ortiz. 1999. Capacidad y Riesgos de Actividades Forestales en el Almacenamiento de Carbono y Conservación de Biodiversidad en Fincas Privadas del Area Central de Costa Rica. IV Semana Científica: "Logros de la Investigación para el Nuevo Milenio." 6-9 de abril de 1999. Programa de Investigación, Centro Agronómico Tropical de Investigación y Enseñanza (CATIE), Turrialba, Costa Rica. Pp. 291-294.

Carnevale, N. J. and F. Montagnini. 2002. Facilitating regeneration of secondary forests with the use of mixed and pure plantations of indigenous tree species. *Forest Ecology and Management* 163: 217-227.

Cusack, D. and F. Montagnini. 2004. The Role of Native Species Plantations in Recovery of Understory Diversity in Degraded Pasturelands of Costa Rica. *Forest Ecology and Management* 188: 1-15.

Evans, J. 1999. Planted forests of the wet and dry tropics: Their variety, nature and significance. *New Forests* 17: 25-36.

FAO. 2000. Global Forest Resources Assessment 2000. Main Report. *http://www/fao.org/forestry/fo/fra/main*

FAO. 2001. State of the world's forest. FAO Technical Papers. Food and Agriculture Organization of the United Nations, Rome.

González, E. and R. Fisher. 1994. Growth of native species planted on abandoned pasture land in Costa Rica. *Forest Ecology and Management* 70: 159-167.

Guariguata, M. R., R. Rheingans and F. Montagnini. 1995. Early woody invasion under tree plantations in Costa Rica: Implications for forest restoration. *Restoration Ecology* 3(4): 252-260.

Haggar, J. P., C. B. Briscoe and R. P. Butterfield. 1998. Native species: A resource for the diversification of forestry production in the lowland humid tropics. *Forest Ecology and Management* 106: 195-203.

Horn, N. and F. Montagnini. 1999. Litterfall, litter decomposition and maize bioassay of mulches from four indigenous tree species in mixed and monospecific plantations. *International Tree Crops Journal* 10: 37-50.

Kershnar, R. and F. Montagnini. 1998. Leaf litter decomposition, litterfall and effects of leaf mulches from in mixed and monospecific plantations in Costa Rica. *Journal of Sustainable Forestry* 7(3/4): 95-118.

Lamb, D. 1998. Large scale ecological restoration of degraded tropical forest lands: The potential role of timber plantations. *Restoration Ecology* 6(3): 271-279.

Masera, O., J. F. Garza-Caligaris, M. Kanninen, T. Karjalainen, G. Nabuurs, A. Pussinen, B. J. de Jong and G. M. J. Mohren. 2003. Modelling carbon sequestration in afforestation and forest management projects: The CO2FIX V 2.0 approach. *Ecological Modeling* 164: 177-199.

Montagnini, F. 2001. Strategies for the recovery of degraded ecosystems: Experiences from Latin America. *Interciencia* 26(10): 498-503.

Montagnini, F., E. González, C. Porras, and R. Rheingans. 1995. Mixed and pure forest plantations in the humid neotropics: A comparison of early growth, pest damage and establishment costs. *Commonwealth Forestry Review* 74(4): 306-314.

Montagnini, F. and R. Mendelsohn. 1997. Managing forest fallows: Improving the economics of swidden agriculture. *Ambio* 26(2): 118-123.

Montagnini, F. and C. Porras. 1998. Evaluating the role of plantations as carbon sinks: An example of an integrating approach from the humid tropics. *Environmental Management* 22: 459-470.

Montagnini, F. and P. K. Nair. 2004. Carbon Sequestration: An under-exploited environmental benefit of agroforestry systems. *Agroforestry Systems* 61: 281-295.

Nabuurs, G., J. F. Garza-Caligaris, M. Kanninen, T. Karjalainen, T. Lapvetelainen, J. Liski, O. Masera, G. M. J. Mohren, M. Olguin, A. Pussinen, and M. J. Schelhaas. 2002. CO2FIX V 2.0: Manual of a modelling framework for quantifying carbon sequestration in forest ecosystems and wood products. Wageningen, ALTERRA Report No. 445. 45 p.

Nasi, R., S. Wunder, and J. J. Campos Arce. 2002. Forest ecosystem services: Can they pay our way out of deforestation? A discussion paper prepared for the GEF for the Forestry Roundtable to be held in conjunction with the UNFF II, Costa Rica on March 11, 2002.

Parrotta, J. A. 1992. The role of plantation forests in rehabilitating degraded tropical ecosystems. *Agriculture, Ecosystems and Environment* 41:115-133.

Petit, B. and F. Montagnini. 2004. Growth Equations and Rotation Ages of Ten Native Tree Species in Mixed and Pure Plantations in the Humid Neotropics. *Forest Ecology and Management* 199: 243-257.

Piotto, D., F. Montagnini, L. Ugalde and M. Kanninen. 2003a. Performance of forest plantations in small and medium sized farms in the Atlantic lowlands of Costa Rica. *Forest Ecology and Management* 175: 195-204.

Piotto, D., F. Montagnini, L. Ugalde and M. Kanninen. 2003b. Growth and effects of thinning of mixed and pure plantations with native trees in humid tropical Costa Rica. *Forest Ecology and Management* 177: 427-439.

Piotto, D., F. Montagnini, M. Kanninen, L. Ugalde, E. Viquez. 2004. Forest plantations in Costa Rica and Nicaragua: Performance of species and preferences of farmers. *Journal of Sustainable Forestry* 18(4) 59-77.

Powers, J. S., J. P. Haggar and R. F. Fisher. 1997. The effect of understory composition on understory woody regeneration and species richness in 7-year old plantations in Costa Rica. *Forest Ecology and Management* 99: 43-54.

Shepherd, D. and F. Montagnini. 2001. Carbon Sequestration Potential in Mixed and Pure Tree Plantations in the Humid Tropics. *Journal of Tropical Forest Science* 13(3): 450-459.

Ugalde, L. ED. 1997. Resultados de 10 años de investigación silvicultural del proyecto MADELEÑA en Panamá. Turrialba, C.R., CATIE. 113 p. (Serie Técnica. Informe Técnico n. 293).

Wadsworth, F. H. 1997. Forest Production for Tropical America. United States Department of Agriculture Forest Service. Agriculture Handbook 710. Washington, DC.

Wormald, T. J. 1992. Mixed and pure forest plantations in the tropics and subtropics. FAO Forestry Paper 103. FAO Technical Papers. Food and Agriculture Organization of the United Nations, Rome. 152 pp.

Carbon Sequestration in Rural Communities: Is It Worth the Effort?

Jens B. Aune
Alene T. Alemu
Kamala P. Gautam

SUMMARY. This study focuses on the possibility for income generation through carbon sequestration in land-use related projects under the Clean Development Mechanism (CDM) of the Kyoto protocol. The carbon sequestration potential of agroforestry and forestry projects was studied in Nepal, Uganda and Tanzania. Carbon sequestration in biomass was 0.43 and 2.0 Mg C ha^{-1} $year^{-1}$ in regeneration of natural forests in Tanzania and Nepal, respectively. Carbon sequestration rates in planted forests were 4.0 Mg C ha^{-1} $year^{-1}$ for *Eucalyptus* woodlots in Uganda and 5.9 Mg C ha^{-1} $year^{-1}$ for *Pinus patula* plantations in Tanzania. The economic profitability of these land-use systems was calculated using the Net Present Value (NPV). Based on a carbon price of US $10 per ton (Mg) and an annual interest rate of 10%, the NPV increased by between 4.9% and 6.5% for these systems when the carbon value was added to the timber value and value of non-wood products. This increase in value is probably too small to justify the cost in relation to project

---

Jens B. Aune is Associate Professor in Agroecology, Center for International Environment and Development Studies, Noragric, Agricultural University of Norway, P.O. Box 5003, N-1432 Aas, Norway.

Alene T. Aleme is Lecturer, Mekelle University, Ethiopia.

Kamala P. Gautam is a former MSc student, Center for International Environment and Development Studies, Noragric, Agricultural University of Norway, Aas, Norway.

[Haworth co-indexing entry note]: "Carbon Sequestration in Rural Communities: Is It Worth the Effort?" Aune, Jens B., Alene T. Alemu, and Kamala P. Gautam. Co-published simultaneously in *Journal of Sustainable Forestry* (Food Products Press, an imprint of The Haworth Press, Inc.) Vol. 21, No. 1, 2005, pp. 69-79; and: *Environmental Services of Agroforestry Systems* (ed: Florencia Montagnini) Food Products Press, an imprint of The Haworth Press, Inc., 2005, pp. 69-79. Single or multiple copies of this article are available for a fee from The Haworth Document Delivery Service [1-800- HAWORTH, 9:00 a.m. - 5:00 p.m. (EST). E-mail address: docdelivery@haworthpress.com].

Available online at http://www.haworthpress.com/web/JSF

doi:10.1300/J091v21n01_04

development, establishment of baseline, carbon monitoring costs, assessment of leakage and documentation of the effect on sustainable development.

KEYWORDS. Carbon sequestration, miombo woodlands, woodlots, plantations, Net Present Value

INTRODUCTION

Climate change is an issue of global concern and has important economic and social consequences. Estimates indicate that temperatures will increase 1.5°C to 6°C in the period from 1990 to 2100 due to emissions of greenhouse gases, and that more extreme weather events are likely to occur (IPPC, 2001). At the same time, poverty is widespread and the World Food Program estimates that 38 billion people in Africa suffer from malnutrition (World Food Program, 2002). For these reasons, it is of interest to identify interventions that can simultaneously contribute to poverty eradication and sequester greenhouse gases in soil and vegetation. The Clean Development Mechanism (CDM) of the Kyoto Protocol may assume such a role. The CDM has been developed with the objective of assisting industrialized countries comply with their quantified emissions limitation and to promote sustainable development in developing countries. Industrialized countries consider CDM as a potential source for low-cost emission credits, while developing countries hope it may attract new and additional investment for sustainable development. CDM projects need to document that the emission reductions are real (additional) and that the emissions reductions are unlikely to occur without the project. Furthermore, projects need to document that they contribute to sustainable development. Most developing countries lack information about the opportunities, likely costs, benefits and risks associated with carbon projects. Carbon trading is rapidly expanding, now that the World Bank and other institutions have established funds to facilitate the establishment of CDM projects (World Bank, 2004).

The profitability of land-use related CDM projects will depend on the price of carbon in the international market, additional income from the project like the sale of timber and the cost related to carbon monitoring. Estimates of future carbon prices are highly uncertain. Brokers believe the price in the period 2005-2007 will be closer to US $4 per ton (Mg) carbon and the prices are expected to rise after 2007 (Nicholls, 2004). The BioCarbon Fund of the World Bank is expecting to pay about US $1 per Mg carbon upon delivery (BioCarbon Fund, 2004).

There are examples that demonstrate how sink-related CDM projects can promote sustainable development. In Mexico, a CDM project is assisting farmers to switch from swidden agriculture to agroforestry, either by combining crops and timber trees or by enriching fallow lands (Nelson and de Jong, 2003). As much as 60% of carbon revenues have gone directly to farmers to cover establishment costs of new farming systems (Smith and Scherr, 2002). The farmers themselves draw up the plans, reflecting their own needs and capabilities with the help of the trust's representatives.

The objective of this paper is to study the potential for carbon sequestration in relation to agroforestry and forestry projects in developing countries and to assess how much additional income can be generated through carbon sequestration.

MATERIALS AND METHODS

Measurements of stocks of carbon were undertaken in forests, agroforestry plots and open cropland in order to assess the carbon sequestration potential of these land-use systems. The profitability of the different land-use systems was assessed by undertaking household surveys and analyzing cost and income in the different land-use systems.

The study was undertaken at sites in Nepal, Uganda and Tanzania. Annual average rainfall in mm is 2200, 1092, 860 and 1300 for the sites in Inner Tarai (Nepal), Kabale (Uganda), Morogoro (Tanzania) and Sao Hill (Tanzania), respectively.

Biomass and C sequestration were measured in six farms in Nepal (Gauatam, 2002). Different land-use categories were identified at each farm and biomass and economic data were collected for each land-use system. Carbon storage and accumulation was assessed in natural forest, fruit production systems and annual cropping systems in Nepal (Table 1). Biomass of larger trees was assessed in quadrats of 10 by 10 meters. Height of the trees was assessed using an Abneys level and tree diameter was measured. Allometric equations developed in Nepal were used to calculate total tree biomass. Understory biomass, including

TABLE 1. Carbon pools, carbon sequestration rate and Net Present Value (US $) in forest and cropping systems in Nepal.

	Biomass carbon (Mg ha^{-1})	Soil carbon (Mg ha^{-1})	Annual C accum. in biomass (Mg ha^{-1})	Annual C accum. in soil (Mg ha^{-1})	Annual carbon value (Mg ha^{-1})	NPV (US $ ha^{-1})
Naturally grown forest	45.7	53.2	2.0	0.6	2.6	1627
Mango	26.8	47.8	1.0	0.6	1.6	306
Rice + banana	24.5	38.6	0.9	−0.4	0.5	n.a.
Rice	6.0	41.0	0	0	0	3015
Maize	7.3	40.0	0.1	0.1	2	920

bushes, was assessed by destructive sampling in squares of one by one meter. Soil carbon was assessed in the soil layer 0-20 cm by taking five sub-samples from each of the 10 by 10 meter plots. The soil organic matter content was measured using the Walkley-Black method (Hesse, 1971) and converted into soil organic carbon by multiplying by a conversion factor of 0.58.

In Kabale, Uganda, carbon sequestration was assessed in an agro-forestry project (AFRENA) (Alemu, 2003). This was a joint project of the World Agroforestry Center and the Uganda Forestry Research Institute whose purpose was to study the regeneration of degraded land through establishment of woodlots. Trees were planted at a density of 2500 trees ha^{-1}. The trees were harvested four years after planting in 21 sites; five trees were sampled per plot. The trees were cut, dried and weighed after harvest. Five soil samples were taken per plot at 0-20 cm.

The possibility of sequestering carbon in naturally regenerating forests was studied in miombo woodlands in Tanzania. Miombo woodlands occupy 95% of the forest area in Tanzania (Sataye et al., 2001). It is an open forest dominated by the tree species *Brachystegia-Julbernardia* and related associations (Pratt el al., 1966). In Morogoro, Tanzania, the average growth of miombo woodland plots was determined to be 1.45 Mg ha^{-1} year^{-1} (Ek, 1994). The growth of the miombo woodlands in Morogoro was based on measurement from 1977/1978 and 1992/1993 in 22 plots. Biomass was calculated based on diameter at breast height (DBH) and height of tree (H) using the following formula:

$$\text{Biomass} = 0.06\ \text{DBH}^{2.012}\text{H}$$

The second study in Tanzania was based on results from a plantation forest of *Pinus patula* Schlect. & Cham. in Sao Hill (Chamshama and Philip, 1980). Stands that varied in age from 4 to 27 years were measured in 99 different plots and used as a basis for calculating annual increments in biomass. Diameter at breast height and height of trees were measured and volume was calculated based on a volume table. Volume was multiplied by the specific gravity of the wood to calculate total biomass.

In the sites in Tanzania, no soil data were recorded. Harvest cost of timber and cost of processing charcoal were not included in the economic calculation in Tanzania.

The Net Present Value (NPV) for the three countries was calculated using an annual interest rate of 10% and a carbon price of US $10 per Mg of carbon. The carbon value was added in the last year of the calculation period.

The results of the forestry and agroforestry systems presented in this study are useful to demonstrate the benefits that can be generated, should these projects be registered under CDM.

RESULTS

Results are presented from one site in Nepal, one site in Uganda and two sites in Tanzania. These sites represent forestry, agroforestry and annual cropping systems. Data are presented for carbon storage, carbon accumulation and NPV with and without the value of the carbon credit.

Carbon Sequestration in Forest and Cropping Land in Nepal

In the sites in Nepal, the natural forest consisted of more than five different tree species. The carbon accumulation rates are based on a 20-year period and the forest was established from natural regeneration (Gautam, 2002). The production systems differed widely with regard to the rate of carbon sequestration. The natural forest system sequestered 2.6 Mg C ha^{-1} $year^{-1}$,whereas the rice system was in equilibrium with regard to the carbon balance. As seen in Table 1, the amount of carbon stored in natural forest was almost two times greater than that of the mango plantation and the rice/banana system. The amount of carbon accumulation in maize and rice monocropping is negligible.

The study from Nepal shows that the most profitable systems are not necessarily the systems that sequester the most carbon. The NPV was highest in the rice system, whereas a low NPV was found in the mango system. Important reasons for the low NPV of the mango system were the high establishment costs and low annual incomes. The income from the natural forest is based on the timber value. The NPV of the forest during a 20 year period will increase by 4.9% if the land-users are paid US $10 per Mg of carbon. By using the Goal Seek function in Microsoft Excel, it was calculated that the price of carbon would need to rise to US $172 if the NPV of the natural forest is to be equal to NPV of the rice production system. It is, therefore, very unlikely that rice will be replaced by natural forest in Nepal.

Carbon Sequestration in Uganda

The results of the study in Kabale, in southwestern Uganda showed that the woodlots were more economically attractive than maize monocropping and can sequester considerable amounts of carbon (Table 2) (Alemu, 2003). Poles for construction are the major economic output from the woodlots of *Eucalyptus camaldulensis* Dehnh and *Alnus acuminata* Kunth. The amount of carbon sequestered in woodlots of *Eucalyptus* and *Alnus* was 6.8 and 5.9 Mg C ha^{-1} year^{-1}, respectively. In Kabale, farmers may therefore change to agroforestry even without considering the carbon sequestration value. If farmers are paid in addition US $10 per Mg of carbon sequestered, this will in-

TABLE 2. Carbon sequestration and economic return in maize and agroforestry systems in Kabale, Uganda.

	Annual C accumulation in biomass (Mg ha^{-1})	Annual C in soil as compared to maize monocropping (Mg ha^{-1})	Total annual C accumulation (Mg ha^{-1})	NPV (US $ ha^{-1})
Eucalyptus	4.0	2.8	6.8	2589
Alnus	3.9	2.0	5.9	3184
Maize	0	0	0	812

crease NPV by 6.1% for *Eucalyptus* and 6.5% for *Alnus*. The discounted value of carbon sequestration could pay for 24% and 29% of the investment costs for planting *Alnus* and *Eucalyptus*, respectively. Improvement in soil fertility as a result of the woodlots is not accounted for in the economic analysis.

Carbon Sequestration in Natural Woodlands and Plantation Forests in Tanzania

Carbon sequestration in regenerating miombo woodlands in Tanzania was found to be 0.43 Mg C ha^{-1} $year^{-1}$ after adjusting for wood-specific gravity and the carbon content of the wood (Table 3). The NPV was calculated based for both with and without a carbon premium. It was based on the assumption that one cubic meter of wood corresponds to 368 kg of charcoal, with an average price of 96 Tanzanian shilling per kg charcoal (Zahabu, 2000). The NPV increased by 4.4% when the carbon value was added.

Average biomass accumulation in plantation forests of *Pinus patula* in Sao Hill during a 28-year period was 11.72 Mg ha^{-1} $year^{-1}$ and the corresponding carbon accumulation was 5.86 Mg carbon ha^{-1} $year^{-1}$ (Chamshama and Philip, 1980) (Table 3). The cost of establishment of plantation was set to US $225 ha^{-1} (Sathaye et al., 2001) and the wood price was equal to US $34.9 per cubic meter timber (United Republic of Tanzania, 2001). The NPV increased by 6.8% when including the carbon value.

DISCUSSION

The amount of carbon sequestered in these studies shows that agroforestry and forestry systems could sequester between 0.43 and 6.8 Mg

TABLE 3. Carbon pools, carbon sequestration rate and Net Present Value for the studied sites in Tanzania (data based on Ek 1994, Chamshama and Philip 1980).

	Biomass carbon (Mg ha^{-1})	Annual C accum. in biomass (Mg ha^{-1})	NPV (US $ ha^{-1})
Regrowth miombo woodlands, Morogoro	7.0	0.43	1928
Pinus patula plantation, Sao Hill (27 years)	149	5.86	24727

C ha^{-1} $year^{-1}$ in soil and vegetation, respectively. The results from Nepal and Uganda indicate that the major portion is sequestered in the tree and crop biomass. These results are in line with studies that showed that forest and agroforestry systems can sequester 0.5 to 5 Mg C ha^{-1} $year^{-1}$ in biomass, and that forest soils can sequester between 0.5 to 2.0 Mg C ha^{-1} per $year^{-1}$ (Niles et al., 2002; Dixon et al., 1994). The carbon sequestration rate for miombo woodlands presented in this study was below the carbon sequestration rates found in the Zambian copper belt (Chidumayao, 1990), but equal to the carbon sequestration rate found in the Northern Province of Zambia (Stromgaard, 1985). The carbon sequestration rate in the *Eucalyptus* fallow was higher than in an *Eucalyptus* fallow in the dry tropics of Cameroon (Harmand et al., 2004). In Uganda, the carbon biomass accumulation was 4.0 Mg C ha^{-1} $year^{-1}$, as compared to 1.7 Mg C ha^{-1} $year^{-1}$ in Cameroon.

Our study showed that the land-use systems that sequestered the most carbon were not always the systems that were the most economically profitable. In some cases, there seems to be a win-win relationship between carbon sequestration and profitability, in other cases the inverse is true. Based on a carbon price of US $10 per Mg and an annual interest rate of 10%, the NPV increased by 4.9% and 6.5% for the systems considered when the carbon value was added. This increase is probably too small to justify the cost in relation to project development, carbon monitoring costs, and documentation of the effect on sustainable development. There is an economy of scale related to CDM projects that will favor projects that cover large areas. Monitoring costs will also most likely be lower in plantation forests than in agroforestry systems and in natural reforestation because of more uniform plant stands. More simplified processes for registration, however, have been developed for smaller CDM projects (UNFCC, 2004).

The additional income that can be generated from registering these projects under CDM are probably overestimated because the costs related to monitoring and registration as a CDM project are not included. The price of carbon used in this study is also higher than the current carbon price. According to the current CDM regulations, all carbon in vegetation is considered released to the atmosphere after harvest. This should be adjusted for in the calculation, but was not done in this study. However, carbon in vegetation can also be considered as a temporary storage of carbon until more permanent solutions for carbon storage and reduction are found. Use of temporary credits that expire after each commitment period has been discussed for land-use related projects by the eight Conference of the Parties of the United Nations

Framework Convention on Climate Change (UNFCCC) (Dutschke and Schlamadinger, 2003; Pew Center, 2002).

Agroforestry and forestry projects can contribute to income generation, poverty reduction and environmental preservation, but it is not clear whether CDM will play a vital role in relation to the development of agroforestry and forestry projects. The carbon benefit as compared to the other benefits will be determined partly by the ratio between output prices of timber/non-timber forest products and the price of carbon. This will favor carbon sequestration in more remote areas where the transport costs of timber and other forest products will be higher. In areas with good infrastructure and good market access, the timber price will be higher and the carbon value of the forest as compared to the timber value will be low.

The capital market is not functioning well in many developing countries and CDM projects may provide funds that can be used to cover the investment cost in relation to CDM projects. For the agroforestry project in Uganda, it was shown that carbon income from a potential CDM project could cover 24% to 29% of the investment. This can be a valuable contribution, but it is still quite modest.

Carbon sequestration projects under CDM can be a potential income source for rural communities in the future. Some projects have been initiated, but it remains a question if developing countries should engage in CDM projects at a large scale. The future price of carbon will be a major factor determining the size of the carbon market. The future carbon price is uncertain because it depends on factors such as future predictions for climate change, the number of countries with obligations to reduce emissions, the cost of reducing greenhouse gas emissions in other sectors of the economy, and new innovations regarding the use of fossil and renewable energy (Chen, 2003). Carbon sequestration projects in the land-use sector must be able to offer carbon credits at a competitive price as compared to the other sectors of the economy. At the present carbon price of about US $4 per Mg carbon, the development of carbon sequestration projects is not very attractive. The other benefits from land-use related projects such as timber, fruits and fodder by far outweigh the value of the carbon credit. The carbon benefit must be seen as an additional value. However, does this additional small income justify registering the project as a CDM project? We believe that in many cases the answer will be negative due to the considerable costs of the registration process, measuring of carbon during a project's life time, assessing leakage and validation. However, in some cases development of land-use related CDM projects can still be warranted. This can be the case if the project can sequester high amounts of carbon (beyond 10 Mg C ha^{-1} $year^{-1}$), if carbon sequestration can be measured at low costs, and if the project covers a large area. The benefit farmers can reap from carbon sequestration projects will also depend on when in the project period the carbon payment is released. Farmers will be able to benefit much more if the carbon payment is released early in the project period, particularly if the interest rate is high.

Land-use related CDM projects are long-term in nature and will require a relatively stable political environment, well functioning national and local institutions, and trained manpower. Many of the poorest countries in Asia and Africa do not offer such conditions, and it is a paradox that the countries that are in most need of the income that CDM projects can generate are probably the least prepared to host such projects. Economical, institutional, political and ecological factors will therefore have to be assessed on a case by case basis before initiating the process of development of new land-use related CDM projects.

REFERENCES

Alemu, A.T. 2003. Carbon sequestration in agroforestry and open cropland at the Kigezi highlands, Southwest, Uganda. Centre for International Environment and Development Studies, Noragric, MSc Thesis, Agricultural University of Norway, 61 pp.

BioCarbon Fund. 2004. Retrieved August 30, 2004 from *http://carbonfinance.org/biocarbon/router.cfm?Page=Projects#2*

Chamshama, S.A.O. and M. Philip. 1980. Thinning *Pinus patula* plantations at Sao Hill, Southern Tanzania. Division of Forestry, Faculty of Agriculture, Forestry and Veterinary Science, University of Dar es Salaam, Record No. 13, 16 pp.

Chen, W.Y. 2003. Carbon quota price and CDM potentials after Marrakesh. *Energy Policy* 31: 709-719.

Chidumayo, E.N. 1990. Above ground biomass structure and productivity in a Zambezian woodland. *Forest Ecology and Management* 36: 33-46.

Dutschke, M. and B. Schlamadinger. 2003. Practical issues concerning temporary carbon credits in the CDM. HWWA Discussion Paper 227. Hamburg Institute of International Economics, 12 pp.

Ek, T.M. 1994. Biomass structure in miombo woodland and semievergreen forest. MSc Thesis, Agricultural University of Norway, 53 pp.

Gautam, K.R. 2002. Carbon sequestration in agroforestry and annual cropping system in Inner Tarai, Central Nepal. Centre for International Environment and Development Studies, Noragric MSc Thesis, Agricultural University of Norway, 64 pp.

Harmand, J.M., C.F. Njiti, F. Bernhard-Reversat, and H. Puig. 2004. Aboveground and belowground biomass, productivity and nutrient accumulation in tree improved fallows in the dry tropics of Cameroon. *Forest Ecology and Management* 188: 249-265.

Hesse, P.R. 1971. A text book of soil chemical analysis, John Murray, London, 520 pp.

IPPC. 2001. Climate change 2001. Climate change 2001: Synthesis Report. Summary for policymakers. An assessment of the Intergovernmental Panel on climate change. Retrieved August 30, 2004 from *http://www.ipcc.ch/pub/un/syreng/spm.pdf*

Nelson, K.C. and B.H.J. de Jong. 2003. Making global initiatives local realities: Carbon mitigation projects in Chiapas, Mexico. *Global Environmental Change* 13: 19-30.

Nicholls, M. 2004. Carbon trading: Making a market. Retrieved June 19, 2004 from *http://www.environmental-center.com/resulteacharticle.asp?codi=3086*

Pew Center. 2002. Conference of the parties (COP 8). Climate talks in New Delhi. Retrieved September 15, 2004 from *http://www.pewclimate.org/what_s_being_done/in_the_world/cop_8_india/index.cfm*

Sathaye, J.A., W.R. Makundi, K. Andrasko. 2001. Carbon mitigation potential and costs of forestry options in Brazil, China, India, Indonesia, Mexico, the Philippines and Tanzania. *Mitigation and Adoption Strategies of Global Change* 6: 185-211.

Smith, J. and S.J. Scherr. 2002. Forest carbon and local livelihood. Assessment of opportunities and policy recommendations. CIFOR Occasional Paper No. 17, Jakarta, 56 pp.

Stromgaard, P. 1985. Biomass, growth and burning of woodland in a shifting cultivation area of south Central Africa. *Forest Ecology and Management* 12: 163-178.

UNFCC. 2004. Simplified modalities and procedures for small-scale Clean Development Mechanism project activities. Retrieved August 30, 2004 from *http://cdm.unfccc.int/pac/Reference/Documents/AnnexII/English/annexII.pdf*

United Republic of Tanzania. 2001. The Forest Ordinance Cap. 389. Government Printer.

World Bank. 2004. Carbon finance at the World Bank. Retrieved August 4, 2004 from *http://carbonfinance.org/*

World Food Program. 2002. Africa hunger alert campaign begins. Retrieved August 30, 2004 from *http://www.wfp.org/index.asp?section=3*

Zahabu, E. 2000. Impact of charcoal extraction to the forest resources of Tanzania: The case study of Kitulangalo area. Unpublished MSc Dissertation, Sokoine University of Agriculture, Morogoro.

Shade Coffee Agro-Ecosystems in Mexico: A Synopsis of the Environmental Services and Socio-Economic Considerations

Sarah Davidson

SUMMARY. Coffee-growing ecosystems have significant environmental benefits and social importance. A review of the literature on ecosystem services, especially for biodiversity, is presented on Mexican coffee cultivation. This study describes the characteristics of the five main coffee cultivation systems in Mexico. Coffee farms can be classified on a continuum according to the extent of shade that is incorporated in cultivation systems and how well they represent traditional coffee farms: traditional 'rustic' or 'mountain' coffee gardens, traditional poly-cultures, commercial poly-cultures, shaded monoculture coffee systems and unshaded monoculture crops. After assessing the importance of the presence or absence of trees for each system, this paper examines the ecosystem services' potential of the different systems and the implications of these services for local and global populations. Finally, the potential use of certification for shade coffee is assessed. Over the last few years, it has become clear that a rigorous shade-certification system

---

Sarah Davidson is affiliated with the WWF Macroeconomics for Sustainable Development Program Office, WWF, Washington, DC, USA.

Address correspondence to: Sarah Davidson, 1356 Madison Street, NW, Washington, DC 20011 USA (E-mail: sarah.davidson@aya.yale.edu).

The author thanks Florencia Montagnini and an anonymous reviewer for comments on this paper.

[Haworth co-indexing entry note]: "Shade Coffee Agro-Ecosystems in Mexico: A Synopsis of the Environmental Services and Socio-Economic Considerations." Davidson, Sarah. Co-published simultaneously in *Journal of Sustainable Forestry* (Food Products Press, an imprint of The Haworth Press, Inc.) Vol. 21, No. 1, 2005, pp. 81-95; and: *Environmental Services of Agroforestry Systems* (ed: Florencia Montagnini) Food Products Press, an imprint of The Haworth Press, Inc., 2005, pp. 81-95. Single or multiple copies of this article are available for a fee from The Haworth Document Delivery Service [1-800-HAWORTH, 9:00 a.m. - 5:00 p.m. (EST). E-mail address: docdelivery@haworthpress.com].

Available online at http://www.haworthpress.com/web/JSF

doi:10.1300/J091v21n01_05

is important, but that its promotion must keep abreast of the ecological diversity and farmers' realities.

KEYWORDS. Biodiversity, certification, industrial agriculture, international markets, traditional knowledge

INTRODUCTION

Coffee is one of the major crops that has left a "major ecological footprint in terms of reduced biodiversity and increased soil and water degradation" in Latin America (Muschler, 2002). Latin America contributes 80 percent of coffee on the global market (Baffes et al., 2005). In addition, some types of traditional coffee growing systems, such as shaded agroforests can maintain landscape biodiversity and decrease market risks by diversifying production. Coffee plantations, usually of the *arabica* bean, require certain humidity and thermal characteristics, and in Mexico are often found between 600 and 1200 meters above sea level, and often located on central and southern coastal slopes. In Mexico, the state of Chiapas contains the largest percent of coffee producers, about 27% (Ridler, 1997), Veracruz produces approximately 23%, Oaxaca contributes just over 17% and Puebla produces about 10% (Nestel, 1995). Coffee plantations are also located in Guerrero, Hidalgo, San Luis Potosí, Nayarit, Jalisco, Tabasco, Colima, Michoacan and Queretaro (Nestel, 1995). In 1995, Mexico was the world's fifth largest coffee producer, but Brazil, Vietnam and Colombia continue to far outweigh Mexico's potential influence in international markets (Nestel, 1995). Together, Brazil, Vietnam and Colombia constitute about 48 percent of global coffee production (Baffes et al., 2005). Environmental services of coffee-growing ecosystems can include carbon sequestration and biodiversity conservation. In this article, I will examine the varying types of coffee cultivation, their origins and the respective environmental and social services of each. This understanding provides a foundation to begin to recognize the web of complex linkages in coffee ecosystems–from the local to the global, and from the social to the ecological.

COFFEE GROWING SYSTEMS IN MEXICO

By the late 19th century, coffee was already cultivated in Mexico. Between 1970 and 2001, the area under cultivation jumped from 356, 253 to 703,341 hectares (ha) (Bray et al., 2002). Coffee farms can be classified on a gradient according to the extent of shade that is incorporated in cultivation systems and how well they represent traditional coffee farms: traditional 'rustic' or 'mountain' coffee gardens, traditional 'poly-cultures,' commercial 'poly-cultures,' shaded monoculture coffee systems and unshaded monoculture crops characterize Mexico's coffee cultivation classes (Moguel and Toledo, 1999).

Starting from the most representative of traditional farms, Moguel and Toledo (1999) classify 'traditional rustic or mountain systems' by four main characteristics. First, 'traditional' cultivation replaces the plants that would otherwise grow beneath tropical forests and thus removes only the lower strata of forest while retaining native trees. Secondly, these systems often exist in isolated regions. Thirdly, 'traditional' cultivation is usually tended by indigenous communities in Mexico. Producers who own up to two hectares make up 69 percent of Mexican coffee producers, and 60 percent of producers are indigenous (Boot et al., 2003). A fourth characteristic of this variety of cultivation is described by its low management intensity and reliance on few inputs from outside the natural system. 'Rustic' systems may not produce as much coffee on a per hectare basis nor in terms of gross quantity. However, total production, including food, fodder and fuelwood needs to be incorporated into a more complete, holistic calculation of total yield.

The next category on the coffee continuum is the traditional 'poly-culture' system, also called coffee gardens (Moguel and Toledo, 1999). These are shaded coffee plantations like the rustic mountain system where coffee is planted under native forest cover. This system requires a certain amount of technical knowledge about managing native and introduced species and on manipulating species so certain species thrive and others disappear. Farmers will manipulate wild and domesticated varieties of tree-like, shrub and herbaceous species. In these systems, farms "reach maximum vegetational and architectural complexity and the highest 'useful diversity' " (Moguel and Toledo, 1999) and provide for subsistence and commercial needs.

Commercial 'poly-culture' systems follow, totally replacing the original forest canopy trees with shade trees good for coffee, often *Inga* species. Because commercial ventures have more access to the capital necessary for converting to and maintaining this kind of system, 'traditional' farmers would likely require external aid if this system were to be implemented. Unless non-governmental organizations (NGOs) fulfill this need, 'traditional' farmers will not likely convert their farms. Commercial systems are often mixed with other market-oriented crops, such as citrus and banana trees. More agrochemical use is required than on 'traditional' farms (Clay, 2004). All production is intended to be market-oriented, whereas 'traditional' systems yield a variety of products which have primarily subsistence uses. Also, paid labor demands are higher on non-traditional farms and owners rely on seasonal labor (Lewin et al., 2004). This system requires engagement with a cash economy to an extent on which many small-scale farmers may not be able to compete, unless they can form cooperatives, or network with NGOs that can provide financing to mitigate the economies of scale that provide a competitive advantage for large-scale producers.

Two modern systems, the shaded and unshaded monocultures, were introduced by the Instituto Mexicano del Café (Mexican Coffee Institute) (INMECAFE) in the early 1970s (Nestel, 1995). The shaded monoculture represents the trend in Mexico to convert traditional coffee agroforestry systems to larger-scale, commercial shade plantations. This conversion has led to a trend where *Inga* species replace the multitude of native species, leading to a loss of biodiversity and other ecosystem services (Peeters et al., 2003). The shaded monoculture type of coffee cultivation requires a monospecific coffee plantation under a monospecific canopy: almost solely *Inga* species are used for shade because of their nitrogen fixing capacity and shade potential. In these systems, producers often choose to use agrochemicals and rely on external labor. In the past, coffee was grown on large landholdings averaging 500 ha or more; now, over half of the world's coffee is grown on farms 5 ha or smaller (Clay, 2004). This is in part due to labor costs: smallholders can replace costly external labor with unpaid family labor if the plot is of a manageable size (Clay, 2004). Producers who have high labor costs and large farms, and that rely on seasonal labor, are very sensitive to changing coffee prices (Lewin et al., 2004).

Unshaded monocultures maintain no tree cover, and coffee bushes are grown under direct sunlight. This system relies entirely on high inputs of agrochemicals and intensive work year-round. Of all the coffee systems described, unshaded monocultures are thought to have the highest yields, at least in the short-term. However, over the long-term, soil and water degradation due to intensive chemical reliance may reduce productivity level and cause soil erosion. Also, the health of laborers is a concern when dealing with chemicals because of the difficulty of taking adequate protective measures and the possible lack of knowledge concerning chemical hazards. Herbicides used to clean fields of all vegetation other than coffee result in soil exposure and erosion (Clay, 2004). Pesticides have caused chemical poisonings in workers, and in the past organochlorines were applied until their environmental persistence and human health hazards were well documented (Clay, 2004). These chemical inputs are expensive and often imported; the ability of local producers to afford them in part depends on the local currency values remaining high. The imported inputs which are necessary for the technified fields can become too expensive compared to the market value of unprocessed coffee (Clay, 2004). 'Traditional' systems, which have subsistence products integrated within coffee production, reduce the risk associated with boom-bust market cycles that are especially pertinent for Mexico's coffee niche of high quality, more expensive coffee.

According to Beer (1987), shade is the most significant characteristic for the ecology and economy of coffee production, and Moguel and Toledo (1999) note that shade also maintains regional landscape equilibrium. Despite the advantages of shaded coffee farms in retaining some level of traditional management and structure, much of the coffee producing land has converted to intensive, unshaded monoculture. According to Moguel and Toledo's survey (1999), which covered about half of Mexico's territory, 11% of this sampled area had converted its coffee area to intensive, unshaded monoculture; 42% remained shaded monoculture; 10% became commercial poly-cultures; and 39% was traditional coffee poly-culture within original forest. These findings are similar to Nestel's (1995) suggestion that 30% of Mexico's coffee regions changed from highly diversified agroforestry to shaded and unshaded monocultures during the 1980s.

ENVIRONMENTAL SERVICES OF AGROFORESTRY COFFEE SYSTEMS

Benefits of shade–Leguminous shade trees increase the quantity of nitrogen in coffee ecosystems due to their relationship with N-fixing bacteria in their roots (Nestel, 1995). Common shade trees include those of the *Inga* and *Gliricidia* species, such as *Inga jinicuil*, *Inga vera* and *Gliricidia sepium*. In addition to N-fixation by leguminous trees, the litter fall provides important nutrients, stabilizes soil and keeps roots in place (Nestel, 1995). In sun monocultures, more solar radiation reaches the ground, which increases soil temperature, decreases the amount of water and soil micro-organisms, and can result in soil degradation (Nestel, 1995). In contrast, shade trees can protect against extreme microclimate changes, provide nitrogen and contribute sometimes as a form of integrated pest management (Beer et al., 1997).

Shade, in addition to altitude, is attributed to increased coffee quality. Muschler (2001) conducted a study in Costa Rica showing that shade-grown coffee generally had better body and acidity than unshaded plantations. Muschler (2001) found that in low-altitude coffee farms, shade promotes 'slower and more balanced filling,' heavier fruit and larger beans, as well as a more uniform ripening of the berries (*Catimur* and *Caturra* varieties of *arabica*). These characteristics improve essential coffee qualities, such as acidity, body and aroma. Shade also reduced the stress on coffee plants by improving microclimate conditions and decreasing nutritional imbalances (Muschler and Somarriba, 1997).

Problems related to shade–A misperception about the role of shade trees in coffee has persisted, linking lower yields with shaded coffee. However, when assessing productivity of shaded systems, research has focused on coffee yields and ignored the non-coffee products (Gobbi, 2000). This misperception has led to much agronomic research and technical services to link increased yields with decreasing shade tree density (Gobbi, 2000). The viability of small coffee farm holdings may not seem competitive if compared with large coffee farms on variables at which larger, single production systems fare better. For instance, although a large, 'modern' farm may have higher coffee yields in a year, a small, 'traditional' farm will yield a wider variety of products. 'Traditional' farms also can maximize the value of their harvest by improving the quality of their product, having it recognized by a third party, and selling it for a higher per unit price.

Another possible misperception of shaded coffee links the production system to the higher occurrence of certain types of fungal diseases than monocultures. Soto-Pinto et al. (2002) found that the incidence of two common coffee pests, the insect Berry borer (*Hypothenemus hampei* Ferr) and the fungal disease Leaf rust (*Hemileiavastatrix* Berk & Br.), were lower in their study than in previous cases that have been used to illustrate the link between shade and insect or disease occurrence. There was no correlation between the incidence of Berry borer, and there was a negative correlation between the number of shade strata and the incidence of Leaf rust (Soto-Pinto et al., 2002). These researchers concluded that the percent of shade cover, light, coffee density, aspect, stand age, basal area and yields were not correlated to the presence of pest, disease and weeds (Soto-Pinto et al., 2002). Additionally, the commonly held view that weeds cause problems with coffee production is not completely true (Soto-Pinto et al., 2002). Weeds have been considered to be pests in multiple studies around the world; however, local uses of weeds abound and they are sometimes used by indigenous people for their needs (Altieri, 1988; Morse and Buhler, 1997). Clearly, more research is needed to determine correlations among shade and pests and to analyze the advantages and disadvantages of chemical versus organic inputs.

Carbon sequestration–The higher potential for carbon sequestration in coffee agroforestry systems than in monocultures may be one avenue to provide workers engaged in coffee agroforestry systems with financial compensation for providing an ecosystem service that otherwise is not captured in the coffee market. In a study by Dejong et al. (1995), two coffee plantations in Chiapas were assessed for carbon sequestering potential. The farmers would plant trees in their coffee plantation, fallows and pasture lands. The scientists estimated that the total potential carbon sequestration ranged from 46.7 to 236.7 tons of carbon/hectare. Net income could range from \$500-1000 ha^{-1}, depending on carbon's value and how much would be paid to farmers (Dejong et al., 1995). The main obstacles to establishing carbon sequestration markets include the social and political constraints of long-term investments by farmers in a risky activity, viability of community control over projects, selection of appropriate trees and linking small farmers to external carbon markets.

Conservation of biodiversity–Moguel and Toledo (1999) reviewed the literature for quantitative data on flora and fauna in Mexican shaded and unshaded coffee systems. They found information for five main categories: plants, arthropods, birds, amphibians and reptiles. A total of 90-120 plant species were found in traditional coffee fields categorized by a complex vegetation structure containing herbs, shrubs and three strata of trees. Ethnobotanical studies they reviewed showed that indigenous management systems contribute species for human consumption and markets (Molino 1986 in Moguel and Toledo, 1999). Arthropod diversity also seems to be high. A site in Chiapas contained arthropods representing 258 families (Ibarra-Nunez 1990 in Moguel and Toledo, 1999). Traditional coffee agro-ecosystems have "high arthropod diversity that, at least on a tree-by-tree basis, is roughly comparable to the diversity found in tropical forest tree canopies" (Perfecto et al., 1997). Shifting traditional coffee systems to coffee monocultures, however, results in a sizeable loss of arthropod diversity (Perfecto et al., 1997). A study by Gobbi (2000) also suggests that shade coffee plantations that resemble the original forest in terms of structural and floristic diversity may help to conserve biodiversity. If shade plantations replace forests and do not reflect the previous landscape's structural diversity, the clearing will affect soil microorganisms and can reduce the number of tree species by 80 percent, as well as, decreasing mammal, reptile, and bat species (Clay, 2004).

Shaded, traditional systems provide bird habitat, according to a review of the relevant literature. Most bird diversity is found in traditional shaded coffee gardens and the least diversity is located in unshaded monoculture (Moguel and Toledo, 1999). Many migratory birds spend the winter in coffee growing areas of Latin America (Migratory Bird Center Fact Sheet, Smithsonian Migratory Bird Center web page). Very little data exist on amphibians and reptiles. Compared to natural forests, even traditional coffee fields have low mammal diversity, although rare and threatened species such as chupamiel (*Tamandua mexicana*), nutria (*Lutra longicaudis*) and viztlacuache (*Coendu mexicanus*), are found there (Moguel and Toledo, 1999). Traditional coffee gardens "mimic forests sufficiently" (Moguel and Toledo, 1999) and can provide refuge for biological diversity.

Traditional Mexican coffee plantations also are being replaced by coffee monocultures shaded by species of *Inga*. The INMECAFE was instrumental in this phenomenon by recommending *Inga* trees for coffee shading. While *Inga* is evergreen and provides continuous shade, few experiments have been conducted in the field to show that *Inga* increases coffee yield (Peeters et al., 2003). In addition, *Inga*-dominated plantations do not maintain the same quality or quantity of the biodiversity and ecological services found in traditional farms (Peeters et al., 2003). A study in traditional and modern *Inga*-shaded plantations in Plan Paredón, Chiapas, showed that although coffee production was similar among the sites, timber production and total tree biomass were significantly higher in the traditional (non-*Inga* shaded) plantations (Peeters et al., 2003).

Gobbi also notes the important social benefits provided by agroforestry coffee, or what he calls "biodiversity friendly" (Gobbi, 2000) coffee. The four most important benefits outlined in Gobbi's study include: (1) better health conditions for farmers and other coffee workers, (2) improved living standards through environmental education, (3) more opportunities for recreation and tourism, and (4) the cultural and aesthetic benefits for farmers and citizens (Gobbi, 2000).

After reviewing the empirical data, Moguel and Toledo conclude that traditional, indigenous coffee agroforestry systems harbor biological diversity and provide high structural complexity (Moguel and Toledo, 1999). There is a possibility that traditional coffee systems can serve as a refuge for wildlife from surrounding areas that have undergone deforestation.

Traditional shade coffee plantations illustrate some of the important characteristics of agroforestry systems in that they provide habitats for a variety of species, serve as perches and nesting sites, provide food resources, and improve local microclimates that are amenable for a wide variety of birds, mammals, arthropods and plants. Coffee agroforestry systems also can serve as sites for seed deposition and germination, and act as buffer zones and as biological corridors (Moguel and Toledo, 1999).

INTERNATIONAL MARKETS, REGIONAL LANDSCAPES AND FARMERS' LIVELIHOODS

Nestel (1995) examined Mexican coffee farming and production in the context of international markets, agricultural landscape and ecology. He studied how international markets determine the behavior of market commodities, and how manipulating land already in agricultural production or by introducing technological innovations, can influence regional ecologies and landscapes. Coffee prices from 1970 through 1980s and related changes in regional landscapes and coffee farm ecologies of Mexico contribute to his analysis of the confluence of international markets and local coffee farms. Interestingly, coffee plantation modernization began in 1970 (Gobbi, 2000). Gobbi noted that the increasing interdependence due to international trade and markets greatly influences the course of local-level decision making (di Castri and Hansen, 1992 cited in Gobbi, 2000). This is a complicated issue, and researchers need to examine this perspective more deeply.

It is interesting to look at how international market forces have altered the types and extent of coffee cultivation in Latin America. Over the last century, increased coffee demand in the United States has resulted in displacement of other crops with coffee. In 1954, a sharp price peak in coffee on the international market provoked many producers to convert their land over to coffee cultivation. In Brazil, virgin jungle was cleared in the Paraná region during the coffee price surge of the 1950s (Streeten and Elson, 1971 cited in Nestel, 1995). During the 1970s, a steady rise in coffee prices provoked modernization schemes to increase yields. Highly diverse polyculture coffee agroecosystems were converted to more homogenous systems with fewer trees. This time is often referred to as the 'coffee bonanza.' Since Mexico ranks only fifth in importance as a coffee-producer worldwide, its strength and competitive advantage rests with its quality rather than with its quantity.

The price of agricultural commodities in international markets positively correlates to the amount of capital and land engaged in the production of export crops in developing countries. Nestel (1995) attributes much of Mexico's changed landscape to coffee demand from the United States by replacing tobacco and cacao farming with coffee farming. Nestel's research concerning the 'coffee-bonanza' of 1970 to 1986 examines how much land was transformed to coffee production. In 1970, coffee contributed about $100 million to the economy, in 1976 coffee produced $400 million and through 1986 coffee contributed approximately $824.5 million (Nestel, 1995). In 1986, this was 5.14% of all of Mexico's exports and about 50% of its agricultural exports (Nestel, 1995). From 1970 to 1982, the amount of land under cultivation increased by over 141,000 hectares in Mexico (Nestel, 1995). Although new technologies, such as those discussed in the next section, were applied, coffee yields did not grow significantly in the most important coffee growing states (see Table 1, Nestel, 1995). Overall, Mexico's coffee yields increased by only .01 tons/ha over 1970 to 1982, suggesting that external inputs such as agrochemicals, and shifting to shaded or unshaded monocultures were not as beneficial as expected (see Table 1, Nestel, 1995).

By the early 1990s, of the 2.8 million hectares of land used to grow coffee in Mexico, Colombia, Central America and the Caribbean, 1.1 million hectares had been converted into modern, technified plantations (Rice and Ward, 1996). In Mexico, 17% of the permanent croplands planted with coffee are modern, reduced-shade farms (Coffee industry study).

TABLE 1. Mexico's coffee yields from 1970 to 1982 (adapted from Nestel, 1995).

State	1970 (tons/ha)	1982 (tons/ha)
Chiapas	0.59	0.53
Veracruz	0.60	0.79
Oaxaca	0.38	0.36

DOMESTIC AND INTERNATIONAL REGULATION VERSUS HOUSEHOLD MANAGEMENT

The International Coffee Organization (ICO) was created in 1962 to set and administer the International Coffee Agreement (ICA), which designated specific export quotas to each country that produced coffee. This agreement was created to maintain a reasonably stable price market (Nestel, 1995). However, the ICA fell apart in 1989 and shortly thereafter, international coffee prices dropped, and Mexico experienced a 'coffee crisis' (Nestel,1995; Ridler, 1983; Rice, 1997).

Socio-economic information is perceived at the household level, where production goals are integrated with consumption needs of the household (Nestel, 1995). However, domestic regulation discouraged certain traditional techniques; for instance, INMECAFE promoted and provided support for intensive use of agrochemicals and equipment and for shifting coffee farms from agroforestry poly-culture systems to coffee monocultures. These trends contributed to Nestel's (1995) findings that 30% of Mexico's coffee cultivation shifted from highly diversified agroforestry systems to monocultures or unshaded systems. This affects household production and consumption, as well as soil and water quality. On a global level, coffee monoculture plantations provide less carbon sequestration capacity and less biodiversity. Also, it is important to note that the additional technological inputs did not significantly increase yields, as expected. While Chiapas received the most technical assistance, only Veracruz's yields increased significantly (Nestel, 1995). Nestel also comes to the interesting conclusion that INMECAFE decisions favored large-scale coffee owners over small- and medium-sized farmers. In terms of reducing rural poverty, small landowners are crucial, especially in Mexico where peasant collectives, ejidos, own 80% of the land in small landholdings.[1] This was a direct consequence of the Mexican Revolution, in which peasants took back land ownership from the large landowners and hacienda control. Thus, favoring large land-holders, as some suggest (e.g., Nestel, 1995) INMECAFE had done, would ignore a significant proportion of the population.

THE INVISIBLE COMPLEXITIES OF CERTIFICATION EFFORTS

While shade, fair trade, and biodiversity-friendly certification attempts are underway in Mexico, it is important to recognize the complexity behind a seemingly simple solution to coffee farming challenges. Due to the amount of coffee consumers, some researchers suggest that international conservation and development institutions and scientists be cautious when promoting coffee production (e.g., Rappole et al., 2003). To be successful, Gobbi (2000) suggests that biodiversity-friendly, certified coffee should provide financial returns that act as incentives for farmers who are retaining forest cover or increasing shade tree density. Conserving a determined number of any species is not equivalent to maintaining levels of species richness and endemism associated with biologically diverse ecosystems. Gobbi (2000) presents several suggestions for cultivating biologically-friendly coffee. First, coffee farms should be under at least 40% shade cover, with even distribution up to 1200 m of elevation. Second, farms must use at least 10 native species at a minimum density of 1.4 trees per hectare. Gobbi (2000) also notes that conservation policies for forest, soil and water are also essential. Hunting and the removal of flora and fauna for commercial interests should not be allowed. In terms of agrochemical use, Gobbi (2000) suggests that only low-toxicity chemicals should be used in accordance with national and international standards. Also, workers should be trained in the use, storage and application of such chemicals (Gobbi, 2000).

Implementation and financing are two main concerns that must be addressed if such programs are to be successful. This may be an inlet for NGOs for coordinating and providing financing or risk management schemes with individual and collective producers.

Many researchers agree that coffee certification strategies are young and need much more work in their coordination and application (Rappole et al., 2003). Many 'certified' labels currently used are not third-party certified. The two internationally acknowledged certification initiatives are the Smithsonian Migratory Bird Center and Rainforest Alliance's certification systems (Lewin et al., 2004). There are many more national and private standards, such as a major European retailer's Utz Kapeh foundation or the monitoring and coordinating system of the European Fair Trade Labeling Organization International (FLO) (Lewin et al., 2004).

CONCLUSIONS

Over the last few years it has become clear that a rigorous shade-certification system is important, but that its promotion must integrate the certification process with concerns for ecological diversity and the role of small-scale farmers. For instance, conservation (organic, shade, biodiversity-friendly) and social justice (fair-trade) themes should be explicitly linked in certification efforts (Philpott and Dietsche, 2003). Otherwise, in light of certification financial incentives, native forest stands face the threat of conversion to shaded systems that do not accurately reflect the native structure or tree density (Rapolle et al., 2003). Increasing extension services and participatory research programs may assist the evolution of coffee farming. For instance, farmers in Chiapas possess traditional knowledge about visible phenomena (e.g., litter decomposition, mineralization of organic matter, significance of macrofauna), as well as information from organic training workshops (Grossman, 2003). More integrated extension services could help farmers learn to better manipulate and manage introduced species. Also, in the current international atmosphere of increased market concentration,[2] NGOs could provide financing, risk management and coordination among small-scale producers that want to access differentiated and certification markets. It is equally important to study farmers' knowledge in order to improve external researchers' understanding of local ecologies and local needs, such as subsistence strategies, risk-averse behavior and the political realities[3] of coffee growing localities.

NOTES

1. Much research has focused on the benefits and fallbacks of collective management, especially in the context of Mexican peasant and indigenous collectives. While this is a very important discussion, in the sake of space I will not delve into the topic here. For further reading on Mexican collectives, see Davidson, S. 2004. "Community responses to international reforms: A case study of San Juan Nuevo Parangaricutiro, Michoacan, Mexico" in *Tropical Resources* 23: 75-82; McDonald, J. 2001. "Reconfiguring the countryside: Power, control, and the (re)organization of farmers in west Mexico" in *Human Organization* 60: 247-258; Purnell, J. 1999. "With all due respect: Popular resistance to the privatization of communal lands nineteenth century Michoacan" in *Latin American Research Review* 34; Brown, P. 1997. "Institutions, inequalities, and the impact of agrarian reform on rural Mexican Communities" in *Human Organization* 56: 102-110.

2. The increasing concentration in coffee markets is evident at several points in the chain: five roasters control over half of global trade while the top five traders make up about half of all global trade (Lewin et al., 2004).

3. See Robert Rice's article (1997) in which he examines coffee's history and its role in the uprising in Chiapas in the early 1990s.

REFERENCES

Baffes, J., B. Lewin, and P. Varangis. 2005. Coffee: Market setting and policies. In M. Ataman and J. Beghin, eds. *Global Agricultural Trade and Developing Countries*. World Bank.

Beer, J., R. Muschler, D. Kass, and E. Somarriba. 1997. Shade management in coffee and cacao plantations. *Agrofor. Systems* 38: 1-3, 139-64.

Bray D., J. Sanchez , and E. Murphy. 2002. Social dimensions of organic coffee production in Mexico: Lessons for ecolabeling initiatives. *Society and Natural Resources* 15: 429-446.

Coffee industry study. Accessed 3/10/04. *http://www.itds.treas.gov/CoffeeIndustry.html*

Dejong, B., G. Montoyagomez, K. Nelson, and L. Soto-Pinto. 1995. Community forest management and carbon sequestration–a feasibility study from Chiapas, Mexico. *Interciencia* 20: 6.

Gobbi, J.A. 2000. Is biodiversity-friendly coffee financially viable? An analysis of five different coffee production systems in western El Salvador. *Ecol. Econ.* 33: 267-81.

Grossman, J.M.. 2003. Exploring farmer knowledge of soil processes in organic coffee systems of Chiapas, Mexico. *Geoderma* 111: 267-87.

Lewin, B., D. Giovannucci, and P. Varangis. 2004. "Coffee Markets: New Paradigms in Global Supply and Demand." Agriculture and Rural Development Discussion Paper 3, 133 pp. The World Bank.

Migratory Bird Center Fact Sheet, Smithsonian Migratory Bird Center. Accessed 3/9/04. *http://nationalzoo.si.edu/ConservationAndScience/MigratoryBirds/Fact_Sheets/deault.cfm?fxsht=1*

Miguel, A. 1990. The ecology and management of insect pests in traditional agroecosystems. In: Ethnobiology: Implications and applications. Proceedings of the first international congress of ethnobiology, Belem, 1988, ed Darrell Posey and William Overal.

Moguel, P., and V. Toledo. 1999. Biodiversity conservation in traditional coffee systems of Mexico. *Cons. Bio.* 13: 1, 11-21.

Morse, S. 1997. Integrated Pest Management. Chapters 1, 2, pp. 7-32, 61-78. Boulder (CO): Lynne Riener.

Muschler, R. 2001. Shade improves coffee quality in a sub-optimal coffee-zone of Costa Rica. *Agrofor. Systems* 51: 2, 131-139.

Muschler, R. 2002. Organic coffee, biodiversity, and agrochemicals: The use of shade trees for low-input coffee production in Central America. Deutscher Tropentag, Witzenhausen, "Challenges to organic farming and sustainable land use in the tropics and subtropics."

Nestel, D. 1995. Coffee in Mexico: International market, agricultural landscape and ecology. *Ecol. Econ.* 15: 165-78.

Peeters, L.Y.K., L. Soto-Pinto, H. Perales., G. Montoya, and M. Ishiki. 2003. Coffee production, timber, and firewood in traditional and *Inga*-shaded plantations in Southern Mexico. *Agric., Ecosystems and Envt.* 95: 481-493.

Perfecto, I., J. Vandermeer, P. Hanson, and V. Cartin. 1997. Arthropod diversity loss and the transformation of a tropical agro-ecosystem. *Biodiv. and Cons.* 6: 935-945.

Philpott, S., and T. Dietsch. 2003. Coffee and conservation: A global context and the value of farmer involvement. *Cons. Bio.* 17: 6, 1844-1846.

Rappole, J., D. King, and J. Rivera. 2003. Coffee and Conservation. *Cons. Bio.* 17: 1, 334-336.

Rappole, J., D. King, and J. Rivera. 2003. Coffee and Conservation III: Reply to Philpott and Dietsch. *Cons. Bio.* 17: 6, 1847-1849.

Rice, R. 1997. The land use patterns and the history of coffee in eastern Chiapas, Mexico. *Agric. and Human Values* 14: 127-143.

Rice, R., and J. Ward. 1996. Coffee, Conservation, and Commerce in the Western Hemisphere: How Individuals and Institutions can Promote Ecologically Sound Farming and Forest Management in Northern Latin America. Washington, DC: *Smithsonian Migratory Bird Council* and Natural Resources Defense Council.

Ridler, N. 1978. Contrived Dualism and Government Policy. *World Devt.* 6: 1, 117-120.

Ridler, N. 1983. Labour force and land distributional effects of agricultural technology: a case study of coffee. *World Devt.* 11: 7, 593-599.

Soto-Pinto, L., I. Perfecto, J. Castillo-Hernandez, and J. Caballero-Nieto. 2000. Shade effects on coffee production at the northern Tzeltal zone of the state of Chiapas, Mexico. *Agric., Ecosystems and Envt* 80: 1-2, 61-69.

Soto-Pinto, L., I. Perfecto, and J. Caballero-Nieto. 2002. Shade over coffee: Its effect on berry borer, leaf rust and spontaneous herbs in Chiapas, Mexico. *Agrofor. Systems* 55: 37-45.

A Review of the Agroforestry Systems of Costa Rica

Alvaro Redondo Brenes

SUMMARY. This review aims to summarize some of the studies in the agroforestry systems of Costa Rica from the 1970s to date. From the 1970s to the mid 1980s, agroforestry systems were characterized by a dominance of the association of indigenous trees from natural regeneration with coffee, cacao, and pastures. However, in the mid-1980s, a number of trials with native species started in Costa Rica, which identified several promising species for reforestation or agroforestry (i.e., *Vochysia ferruginea*, *V. guatemalensis*, *Terminalia amazonia*, *Hyeronima alchorneoides*, and others). In the 1990s to date, there has been a diversification in the types of agroforestry systems used, and an increase in research on these systems (i.e., taungya, improved fallows, multipurpose trees in pastures, alley cropping, and home gardens). Current research not only focuses on the benefits of trees to the systems, but it also studies the proper combinations of crops-trees to attain more profitable yields

---

Alvaro Redondo Brenes is a PhD student at Yale School of Forestry and Environmental Studies, New Haven, 205 Prospect Street, CT 06511 USA (E-mail: aredondo1@costarricense.cr and alvaro.redondo@yale.edu).

The author thanks all the researchers who contributed to this paper with their findings. He also thanks F. Montagnini, M. Fernández, R. Barr, S. Prasad and two anonymous reviewers for their comments on previous drafts of this manuscript.

[Haworth co-indexing entry note]: "A Review of the Agroforestry Systems of Costa Rica." Brenes, Alvaro Redondo. Co-published simultaneously in *Journal of Sustainable Forestry* (Food Products Press, an imprint of The Haworth Press, Inc.) Vol. 21, No. 1, 2005, pp. 97-119; and: *Environmental Services of Agroforestry Systems* (ed: Florencia Montagnini) Food Products Press, an imprint of The Haworth Press, Inc., 2005, pp. 97-119. Single or multiple copies of this article are available for a fee from The Haworth Document Delivery Service [1-800-HAWORTH, 9:00 a.m. - 5:00 p.m. (EST). E-mail address: docdelivery@haworthpress.com].

Available online at http://www.haworthpress.com/web/JSF

doi:10.1300/J091v21n01_06

and to avoid risk, and highlights the environmental services provided by the systems.

KEYWORDS. Reforestation, plantation crop combinations, taungya, improved fallow, silvopastoral systems, multipurpose trees, alley cropping, home gardens, Costa Rica

INTRODUCTION

The use of agricultural land is undergoing rapid changes in response to increasing environmental concerns and external market forces. Thus, the integration of trees into agricultural landuse systems assumes special importance (Muschler and Bonnemann, 1997). Tropical farmers conserve or actively plant trees in their agricultural fields for a variety of reasons: timber production, firewood, fruits and resins, microclimatic protection, soil conservation, and aesthetic reasons (Schaller et al., 2003). "Agroforestry systems have been considered sustainable land use alternatives for the humid tropics because they may imitate characteristics of natural ecosystems–notably those that have beneficial effects on soil properties" (Tornquist et al., 1999: p. 19). Furthermore, Dagang and Nair (2003) mention that due to deforestation and degradation of productive soils, researchers have had to investigate agroforestry as an alternative to traditional approaches to land management. The most important sets of processes in which trees improve soil productivity are mitigating runoff and soil erosion, maintaining soil organic matter and physical properties, increasing nutrient inputs through nitrogen fixation and uptake from deep soil horizons, and promoting more closed nutrient cycling (Young, 1997).

One of the most remarkable features of Costa Rica is its climatic, edaphic and topographic variability. This allows for a high diversity of flora and fauna. These features also are constraints when a monoculture plantation is being planned, due to the complexity of natural forces and biological systems (Fournier, 1981). Until the early 1900s Costa Rica was almost entirely covered by rich and diverse tropical forests. As agriculture expanded, the forests were replaced with a mosaic of pastures, coffee fields and forest fragments (Harvey et al., 1998). Agroforestry systems in Costa Rica have appeared as an alternative for ameliorating deforestation and the problems arising from monoculture production (Fournier, 1981).

Despite the knowledge of the beneficial effects of trees on soils as mentioned above, reforestation and agroforestry projects are often hampered by the lack of information on appropriate species for reforestation (Tilki and Fisher, 1998). In the last three decades, several researchers have studied different alternatives for tree and shrub species suitable for reforestation or for agroforestry systems. In this paper, I review several studies in agroforestry systems that have been conducted in Costa Rica and present their conclusions about which trees and shrubs best fit these systems.

AGROFORESTRY SYSTEMS IN COSTA RICA: 1970S AND 1980S

Fournier (1981) discusses some of the first studies that were conducted in agroforestry systems in Costa Rica in the 1970s and earlier 1980s (Table 1). The agroforestry systems cited by Fournier are the combination of the native species *Cordia alliodora* (Ruiz & Pavón) Oken. with pastures, coffee (*Coffea arabica* L.), and cacao (*Theobroma cacao* L.). Likewise, other tree species are mentioned such as *Alnus acuminata* H.B.K. in pastures and coffee combinations, *Cedrela odorata* L. with coffee and *Bactris gasipaes* Kunth., and the exotic *Cupressus lusitanica* Miller. in agro-silvopastoril systems. Preliminary analysis of these systems suggests that the incorporation of trees in the agricultural landscape would increase the income of farms (Beer et al., 1979; Rosero and Gewald, 1979; Ford, 1979; cited by Fournier, 1981). Fournier (1981) also describes the advantages and disadvantages of these systems, emphasizing their importance for integrated pest management, soil protection, nitrogen fixation and the production of biomass as a source of raw material for energy production and energy development. For these reasons, it is suggested that agroforestry systems should be considered a good alternative for land use in Costa Rica.

These studies were the onset of promising research and implementation of the use of trees not only in agroforestry systems, but also for reforestation in Costa Rica. The Tropical Agricultural Research and Higher Education Center (CATIE) located in Turrialba, Costa Rica, has provided most of the guidelines for the development and implementation of these systems not only in Costa Rica, but also in Central America. Agroforestry as a structural unit of the Center was established at

TABLE 1. Agroforestry systems of greater use in Costa Rica in the 1970s.

Region	Components	Reference
Atlantic Zone	*Cordia alliodora* (Ruiz & Pavón) Oken–*Coffea arabica* L.	(1)
Atlantic Zone	*Cordia alliodora* (Ruiz & Pavón) Oken–Sugarcane (*Saccharum officinarum* L.)	(1)
La Suiza, Turrialba	*Cordia alliodora* (Ruiz & Pavón) Oken–*Coffea arabica* L.–*Erythrina poeppigiana*	(1)
Atlantic Zone	*Cordia alliodora* (Ruiz & Pavón) Oken–*Theobroma cacao* L.	(2)
Atlantic Zone	*Cordia alliodora* (Ruiz & Pavón) Oken–pasture	(2)
Las Nubes, Coronado	*Alnus acuminata* H.B.K.–pasture	(3), (4)
San Antonio, Coronado	*Alnus acuminata* H.B.K.–*Coffea arabica* L	(5)
Santa Clara, San Carlos	*Cedrela odorata* L.–*Coffea arabica* L.	(6)
Tabarcia, Mora	*Cedrela odorata* L.–*Bactris gasipaes* Kunth.–*Zea mays* L.	(7)
San José de la Montaña, Barva, Heredia	*Cupressus lusitanica* Miller.–pastures	(8)
Several locations	Living fences with different combinations	(9)

Adapted from Fournier (1981). Reference: (1) Beer et al. (1979), (2) Rosero and Gewald (1979), (3) Poschen (1979), (4) Combe (1979), (5) Fournier (1979), (6) Ford (1979), (7) Fournier (unpublished data), (8) González et al. (1979), and (9) Sauer (1979).

CATIE in 1976 and has since undergone rapid development and growth (Muschler and Bonnemann, 1997).

EVALUATION OF NATIVE SPECIES FOR REFORESTATION

In Costa Rica, during the 1970s and 1980s, reforestation programs were established using few tree species, especially exotics such as *Cupressus lusitanica*, *Pinus* spp., *Gmelina arborea* L. and *Eucalyptus* spp., as well as the native species *Cordia alliodora* (Moreira and Palma, 1987; cited by González and Fisher, 1994). Although some planted species have shown high adaptability and rates of growth, problems in establishing some tree species have been observed in the degraded soils of abandoned pastures (González and Fisher, 1994). Unfortunately, most of Costa Rican lowlands, especially in the Atlantic and Northern Region are acidic, as well as, degraded due to previous land use (Tilki and Fisher, 1998; Berti, 1999). Research subsequently focused on tree species that could suit these poor soil conditions.

The Organization for Tropical Studies (OTS) initiated research on the growth of native and exotic timber species in degraded pasture lands in 1985 (Butterfield, 1995; 1996). The study was conducted at La Selva Biological Station in the Atlantic lowlands of Costa Rica. A total of 84 species (52 locally indigenous, 15 native to other regions of Costa Rica and 17 exotic to Costa Rica) were established in mixed stands of 8-12 species (Butterfield, 1995; Haggar et al., 1998). In the plots, they evaluated survival, adaptability, growth, form, and susceptibility to diseases to find out which species could be recommended for reforestation programs in the region (González and Fisher, 1994; Butterfield, 1996). As a result of different trials conducted by the researchers, they concluded that native species appeared to be more adapted than exotic species to the local plantation conditions (González and Fisher, 1994; Butterfield, 1995; 1996). The exotic species probably require higher input forestry practices than native species. Some exotic species in their trials showed low survival and declining growth rates over time, which may be related to a combination of poor adaptation to a lack of dry season, compacted soils, inadequate site preparation and early termination of weed control (Haggar et al., 1998: p. 202).

Butterfield (1995) suggested testing 11 out of the 84 species of these trials for their potential for agroforestry systems (Table 2). She recommended a set of species for taungya and alley cropping systems according to the particular characteristics of the species studied. For instance, *Inga edulis* Martius and *Goethalsia meiantha* (J.D.Sm.) Burret, which have high biomass production, good coppicing ability and high foliar N content (that only for *Inga*), are good candidates for alley-cropping systems. Other species, such as *Callophyllum brasiliense* Cambess and *Virola koschnyi* Warb., which are slow-growing, and have small crowns, but are of high timber value, can grow in association with crops for many years. The intercropping offsets high plantation establishment costs (Butterfield, 1995). As mentioned, these species have the potential to be used in agroforestry systems, but should be tested in the long term. The outcomes obtained by the researcher were from only 3 year-old tree plantations (Butterfield, 1995).

The influence of trees on soil properties should be a determining factor in the choice of species for tree-crop combinations or for tree plantations on degraded sites in the humid tropics (Tilki and Fisher, 1998). In

TABLE 2. Potential species for agroforestry use; average survival, diameter at breast height (dbh) and height at 3 years of age. La Selva Biological Station, Sarapiqui, Costa Rica.

Species	Survival (%)	dbh (cm)	Height (m)	Group
Light shade, short rotation				
Jacaranda copaia (Aublet) D. Don	86	11.9	7.8	Native
Ochroma pyramidale (Cav. ex Lam.) Urban	55	16.4	11.6	Native
Terminalia ivorensis A. Chev.	54	13.6	9.5	Exotic
Eucalyptus deglupta Blume	63	9.0	9.6	Exotic
Small crown, medium rotation				
Callophyllum brasiliense Cambess.	76	2.9	3.3	Native
Dipteryx panamensis (Pittier) Record & Mell	83	3.6	4.8	Native
Laetia procera (Poeppig) Eichl.	98	6.9	6.5	Native
Terminalia amazonia (J.F. Gmelin) Exell	70	7.1	6.9	Native
Virola koschnyi Warb	72	2.8	2.6	Native
Alley Cropping (High biomass)				
Goethalsia meiantha (J.D.Sm.) Burret	93	12.0	7.6	Native
Inga edulis Martius	97	10.6	6.3	Exotic

Adapted from Butterfield (1995).

Costa Rica, several studies of effects of trees on soil properties have been conducted in the last two decades (Montagnini et al., 1989; Montagnini and Sancho, 1994a; 1994b; Stanley and Montagnini, 1999; Tornquist et al., 1999; Montagnini, 2000). At La Selva Biological Station, in just two and a half years, soil conditions improved in three native plantations compared to abandoned pastures (Table 3). In the top 15 cm, soil nitrogen and organic matter were higher under the trees than in nearby pasture, with values close to those found in adjacent 20-year-old forests (Montagnini, 2000). However, Tornquist et al. (1999) conclude in their study comparing the effects of agroforestry systems on soil characteristics that agroforestry systems at five years of age do not appear to improve soils. Notwithstanding the differences in the outcomes

TABLE 3. Effects of trees on soil properties such as pH, total carbon (C), total nitrogen (N), and extractable phosphorus (P) in an experimen site in the Atlantic lowlands of Costa Rica. Tree plantations are 2.5 years of age and the secondary forest is 20-year-old.

Tree species	pH	C (%)	N (%)	P ($cmol\ kg^{-1}$)
Stryphnodendron microstachyum Poepp. et Endl.	5.4	3.42	0.23	5.6
Vochysia ferruginea Mart.	5.4	3.76	0.32	7.1
Vochysia guatemalensis D.Sm.	5.3	3.13	0.29	5.2
Hyeronima alchorneoides Fr. Allemao	5.1	2.96	0.22	1.5
Abandoned pasture	5.3	2.73	0.22	4.9
Secondary Forest	5.3	4.33	0.33	3.6

Adapted from Montagnini and Mendelsohn (1997).

obtained in both studies, the former have continued with positive benefits of the trees on soils over 10 years of research (F. Montagnini, pers. com. 15 March 2004; also see Montagnini and Sancho, 1994a; 1994b; Stanley and Montagnini, 1999).

As a result of these trials, species such as *Callophyllum brasiliense*, *Vochysia ferruginea*, *V. guatemalensis*, *Hyeronima alchorneoides*, *Virola koschnyi*, *Terminalia amazonia* (J. F. Gmelin) Exell and *Dipteryx panamensis* (Pittier) Record are part of the reforestation programs in the region (Piotto et al., 2003). Furthermore, most of these species are being used in silvopastoral combinations with beef cattle once the trees reach five years of age, as their canopy is more open and allows for the growth of natural grasses (Montagnini et al., 2003).

In addition to these trials with native species in the northern region of Costa Rica, one project began in the southern region in 1993 (Leopold et al., 2001) and another in the dry region of Guanacaste in 1995 (Piotto et al., 2004). The former is a project that examines soil restoration by planting mixed stands of native hardwoods. Involving up to 41 species, results show sun-loving species growing as much as 3.1 m $year^{-1}$ in height, and exceeding 10 cm dbh (diameter at breast height) in five years. Mixed stands of indigenous species are proposed as an alternative to monocultures, providing a possible source of income for small farmers, stabilizing the soil, and stimulating the restoration of biodiversity (Leopold et al., 2001). Some of the species with better performance are *Schizolobium parahyba* (Vell.) Blake, *Terminalia amazonia*, *Albizia longepedata* (Pittier) Britton, *Platymiscium pinnatum* (Jacq.) Dugand., *Enterolobium cyclocarpum* (Jacq.) Griseb., *Pithecellobium arboreum* (L.) Urb., *Vochysia ferruginea*, and *Brosimum utile* (Kunth) Oken.

The second study was conducted with thirteen native species and one exotic of the dry tropics of Costa Rica in pure and mixed plantations. Plantations of *Tectona grandis* (exotic) seem to be well adapted to the region, and are certainly a commercially interesting alternative (Piotto et al., 2004). However, Piotto et al. (2004) also consider native species in mixed plantations as a good alternative to contribute to sustainable management and to increase the range of benefits. At 68 months of age, the species with the best performance were *Samanea saman* (Jacq.) Merril, *Dalbergia retusa* Hemls., *Astronium graveolens* Jacq., *Swietenia macrophylla* King, *Schizolobium parahyba*, *Terminalia oblonga* (Ruiz & Pav.) Steud., *Anacardium excelsum* (Bert. & Balb. Ex Kunth) Skeels, and *Pseudosamanea guachapele* (Kunth) Harms. The species with the best performance in reforestation programs may be good alternatives for agroforestry systems such as taungya (F. Montagnini, pers. com., 25 September 2004).

AGROFORESTRY SYSTEMS IN COSTA RICA: FROM THE 1990S TO DATE

During the 1970s and earlier 1980s, silvopastoral systems and associations of trees with coffee and cacao seem to be the most generalized agroforestry systems in Costa Rica (Fournier, 1981). Furthermore, the research was a description of the existing systems and practices (Muschler and Bonnemann, 1997). In the last decade, there has been an increase in not only the number of studies of agroforestry systems, but also their breadth (Tables 4, 5, and 6). The systems which will be described here include: plantation crop combinations, taungya, silvopastoral systems, improved fallows, alley cropping, home gardens, and multi-purpose trees in crop lands.

Plantation Crop Combinations

Plantation crop combinations are either integrated multistory, a mixed, dense mixture of plantation crops, mixtures of plantation crops in alter-

TABLE 4. Examples of plantation crop combination systems in Costa Rica in the last two decades.

Location	Components	Reference
Turrialba	Coffee–*Erythrina poeppigiana* (Walp.) O.F. Cook	(1)
Several locations	Coffee and Cacao–several associations	(2)
Several locations	Organic vs. conventional Coffee	(3)
Southern region	Coffee with and without trees	(4)
Juan Viñas, Cartago	Coffee–*Eucalyptus deglupta* Blume	(5)
Perez Zeledón	Coffee–several timber trees	(6)
Central Valley	Coffee–*Erythrina poeppigiana*	(7)
Turrialba	Coffee and Cacao–*Cordia alliodora*, *Erythrina poeppigiana* and *Cynodon plectostachyus* (K. Schum.) Pilger.	(8)
San Carlos-Siquirres	*Bactris gasipaes*–*Cordia alliodora*	(9); (10)
Guápiles	*Bactris gasipaes*–annual crops	(11)

References: (1) Muschler (2001); (2) Somarriba et al. (2001b); (3) Lyngbæk et al. (1997); (4) Mehta and Leuscher (1997); (5) Schaller et al. (2003); (6) Tavares et al. (1999); (7) Babbar and Zak (1994); (8) Fassbender et al. (1991); (9) Schlönvoigt and Schlönvoigt (1999); (10) Montenegro and Bogantes (2001); (11) Leal et al. (2000).

native rows, or plantation crops (i.e., coffee, cacao, coconut, etc.) in other regular arrangements with fruit, fuelwood, or fodder trees (Nair, 1990). Coffee (*Coffea arabica* L.), cacao (*Theobroma cacao* L.) and peach palm or pejibaye (*Bactris gasipaes* H.B.K.) in association with trees are some examples of the principal plantation crop combinations in Costa Rica (Table 4).

Shade trees that are commonly associated with coffee plantations are *Cordia alliodora*, *Erythrina poeppigiana* (Walp.) O.F. Cook, *Eucalyptus deglupta* Blume, *Leucaena leucocephala* (Lam.) De Wit. and *Cedrela odorata* (Fournier et al., 1981; Nygren and Ramirez, 1995; Mehta and Leuschner, 1997; Somarriba et al., 2001a). Other timber species are being used in southern Costa Rica as shade trees such as *Terminalia amazonia*, and *T. ivorensis* A. Chev., as well as some trees from natural regeneration such as *Aspidosperma megalocarpon* Müll., *Lafoensia punicifolia* DC. and *Ocotea tonduzii* Standl. (Tavares et al., 1999). Shade trees associated with cacao plantations are mainly leguminous species, such as *Gliricidia sepium* (Jacq.) Steud., *E. poeppigiana* and

TABLE 5. Examples of Taungya, silvopastoral and improved fallow systems in Costa Rica used in the last two decades.

System/Location	Components	Reference
Taungya		
Talamanca	*Acacia mangium* Willd. and *Cordia alliodora–Zea mays* L., *Zingiber officinale* Roscoe–*Eugenia stipitata* McVaugh	(1)
Talamanca	*Cordia alliodora*–different systems	(2)
Sixaola	*Eucalyptus deglupta* and *Cordia alliodora*–Cassava (*Manihot esculenta* Crantz.) and *Zea mays* L.	(3)
Sarapiqui	Six tree timber species–3 perennial and 2 annual crops	(4)
Improved fallows		
Coto Brus, Puntarenas	Bean (*Phaseolus vulgaris* L.)–*Erythrina poeppigiana*, *Calliandra calothyrsus* Meissner., *Gliricidia sepium* (Jacq.) Kunt ex Walp, and *Inga edulis* Martius	(5)
Sarapiqui	Six native timber trees	(6)
Silvopastoral		
Guápiles	*Acacia mangium* and *Eucalyptus deglupta*–pastures	(7)
Sarapiqui	Native timber trees–pastures	(8)
Monteverde - San Carlos	Remnant trees in pastures	(9); (10)
Turrialba	Protein Banks: *Gliricidia sepium* and *Erythrina poeppigiana*	(11); (12)

Reference: (1) Kapp and Beer (1995); (2) Somarriba et al. (2001a); (3) Schlönvoigt and Beer (2001); (4) Haggar et al. (1999); (5) Keller (1997); (6) Montagnini and Mendelsohn (1997); (7) Andrade et al. (2000); (8) Montagnini et al. (2003); (9) Harvey et al. (1998); (10) Souza de Abreu et al. (2000) (11) Camero (1995); (12) Camero et al. (2001).

Inga edulis; timber tree species, such as *Cordia alliodora*, *Terminalia ivorensis* or *Tabebuia rosea* (Vertol.) DC. (Somarriba et al., 2001a); or more complex systems with plantains (*Musa* spp.) and *C. alliodora* (Ramirez et al., 2001). Peach palm plantations are often associated with *Cordia alliodora* (Schlönvoigt and Schlönvoigt, 1999; Montenegro and Bogantes, 2001).

Research has been conducted by CATIE's researchers on agroforestry systems with coffee and cacao for more than twenty years in Costa Rica (Beer et al., 1979; Fassbender et al., 1991; Beer et al., 1998; Somarriba et al., 2001a; Ramirez et al., 2001). Research evolved in

TABLE 6. Examples of multipurpose trees in crop lands, alley cropping, multipurpose woody hedgerows and home garden systems in Costa Rica.

System/Location	Components	Reference
Multipurpose trees in crop lands		
San Vito, Coto Brus	*Terminalia amazonia* (J.F. Gmelin) Exell–legume trees	(1)
South-western	*Vanilla planifolia* L.–*Erythrina lanceolata* Standl.	(2)
Turrialba	Tomato (*Lycopersicon esculentum* Mill.)–*Erythrina poeppigiana* and *Gliricidia sepium*	(3)
Sarapiqui	Legumes for acid soils	(4)
Atlantic Zone	Tree species for farm boundaries	(5)
Alley cropping and multipurpose woody hedgerows		
Turrialba	Bean–*Erythrina berteroana* Urb. and mulch of several species	(6)
Sarapiqui	*Zea mays* L.–mulch of several species	(7); (8)
Home gardens		
Nicoya	Several species	(9)
Tortuguero	Several species	(10)

Reference: (1) Nichols et al. (2001), (2) Berninger and Salas (2003); (3) Chesney et al. (2000); (4) Tilki and Fisher (1998); (5) Kapp et al. (1997); (6) Tardieu et al. (2000); (7) Montagnini et al. (1993); (8) Byard et al. (1996); (9) Ochoa et al. (1998); (10) Meléndez (1996).

three majors fields: (a) improvement of traditional agroforestry systems with coffee and cacao; (b) nutrient cycling in shaded coffee and cacao; and (c) shade management in cacao plantations (Somarriba et al., 2001a). The results of this research have shown that the relative importance and overall effects of the different interactions between shade trees and coffee/cacao are dependent upon site conditions (soil/climate), component selection (species/varieties/provenances), below-ground and above-ground characteristics of the trees and crops, and management practices (Beer et al., 1998). The major physiological benefits that coffee and cacao receive from shade trees can be placed in two main categories, both related to reduce plant stress:

1. Amelioration of climatic and site conditions through (i) reduction of air and temperature extremes (heat at lower elevations and cold at higher elevations), (ii) reduction of wind speed, (iii) buffering

of humidity and soil moisture availability, (iv) improvement or maintenance of soil fertility, and (v) erosion reduction;
2. Reduction in the quality and quantity of transmitted light, which results in an avoidance of over-bearing, like in coffee, and/or excessive vegetative growth, like in cacao. Shade also reduces nutritional imbalances and dieback (Beer et al., 1998).

Shade tree species associated with these ecosystems contribute to maintain and improve soil fertility (Fassbender et al., 1991; Nygren and Ramirez, 1995; Mehta and Leuschner, 1997; Beer et al., 1998). For instance, soil organic matter (SOM) content may increase with time under the agroforestry systems of coffee and cacao. The presence of shade tree species may also ameliorate soil erosion (Beer et al., 1998). Furthermore, Babbar and Zak (1994) found higher rates of N mineralization in coffee plantations shaded by *E. poeppigiana* (148 kg N ha^{-1} y^{-1}) compared to unshaded plantations (111 kg N ha^{-1} y^{-1}). They concluded that the N cycle is more conservative in shaded plantations than in unshaded plantations.

Shade trees may also affect product quality. Muschler (2001) found that the shade of *E. poeppigiana* substantially improves coffee quality (varieties *Catimor* and *Caturra*) compared to unshaded plantations. Shading offers the following main benefits: (1) higher weight of fresh fruits; (2) larger beans; (3) higher ratings for visual appearance of green and roasted beans; (4) better ratings for acidity (*Catimor* only) and body; and (5) absence of off-flavors.

Another reason for maintaining shade trees in perennial-crop plantations is the additional income provided by their fruit and/or timber. These products may supplement farmers' incomes when coffee and cacao prices are low (Beer et al., 1998), as well as reduce uncertainty and risk to the farmers (Mehta and Leuschner, 1997; Ramirez et al., 2001). Ramirez et al. (2001) evaluated the financial returns, stability, and risk for monoculture plantations of cacao, plantain and *Cordia alliodora* (laurel) compared to their performance in agroforestry systems. They found that the agroforestry systems were better than the corresponding monocultures in terms of expected net present value (NPV) of income and risk at different discount rates (4, 6, 8 and 12%). For instance, the monocultures of cacao, plantain and laurel at 12% of discount rate and 1,111 planting density ha^{-1} had a NPV of \$6,220, \$8,056, and \$225 ha^{-1}, respectively, compared to the combination of 833 planting density ha^{-1} of cacao, 278 planting density ha^{-1} of plantain and 69 planting density ha^{-1} of laurel which had a NPV of US \$10,598 ha^{-1}.

Mehta and Leuschner (1997) conducted a financial analysis comparing monocultures of *Cupressus lusitanica* and coffee with agroforestry systems of coffee in association with *E. poeppigiana*, *Eucalyptus deglupta* and *Leucaena leucocephala* in Southern Costa Rica. They found that agroforestry systems were more financially attractive than either coffee without trees (monoculture) or the cypress plantations alone. The NPV (10% discount rate) depicted the following values (US \$ ha^{-1}): cypress \$100.4, coffee without trees \$1,269.5, coffee/*E. poeppigiana* \$1,486.2, coffee/*E. deglupta* \$1,484 and coffee/*L. leucocephala* \$1,539.0. Growing coffee in combination with trees both lowers cost and diversifies plantation output, thereby reducing risk (Mehta and Leuschner, 1997).

Peach palm, cultivated for either its fruits or its heart, is another perennial crop which is now being associated with different species such as *Cordia alliodora* (Schlönvoigt and Schlönvoigt, 1999; Montenegro and Bogantes, 2001). The preliminary results of the presence of *C. alliodora* depicted some improvement in soil properties (Montenegro and Bogantes, 2001) and that *C. alliodora* grows better when combined with peach palm than alone (Schlönvoigt and Schlönvoigt, 1999), this association did not affect positively the peach palm yields (Montenegro and Bogantes, 2001). In a study to evaluate peach palm growth and productivity in monoculture, and in combination with *Cordia*, the NPV obtained was \$1,950, \$1,751 and \$1,202 ha^{-1} $year^{-1}$ for monoculture of peach palm, peach palm + 90 *Cordia* trees ha^{-1} from natural regeneration, and peach palm + 360 *Cordia* trees ha^{-1} from natural regeneration, respectively (Montenegro and Bogantes, 2001). Further studies should be conducted in order to evaluate this association in the long term.

In other studies, positive effects on yields and NPV in peach palm were found when associated with *Inga edulis*, *Cedrelinga catenaeformis* Ducke, and *Eugenia stipitata* McVaugh (Diaz et al., 1993, cited by Montenegro and Bogantes, 2001). Leal et al. (2000) found a positive effect in young and old stems of peach palm when it was intercropped with maize (*Zea mays* L.) and cowpea [*Vigna unguiculata* (L.) Walp.]. Maize and cowpea also provided a good income for the farmers in the short term, with NPVs of \$334 and \$162 ha^{-1} $year^{-1}$, respectively.

Taungya Systems

Taungya systems are combined stands of woody and agricultural species during early stages of establishment of plantations. Usually their components are plantation forestry species with common agricultural crops (Nair, 1990). This temporary association of annual crops in juvenile tree plantations can reduce or offset the initial costs of reforestation, and hence provide incentives for tree planting on private lands (Schlönvoigt and Beer, 2001).

In Costa Rica some taungya systems were tested with *Cordia alliodora*, *Acacia mangium* Willd. and *Eucalyptus deglupta* as the woody species component (Table 5). In one of these studies, *Cordia* growth was evaluated in pure plantation (tree density of 1,111 ha^{-1} in monoculture) and in line planting (tree density of 400 ha^{-1} in alleys), in combination with cacao (new and old plantations and with plantain), and in a taungya system with the following crops: maize, ginger (*Zingiber officale* Roscoe) and araza (*Eugenia stipitata*) as a perennial fruit tree (Somarriba et al., 2001b). *Cordia* growth varied among different plantation types and was highest in the cacao-*Cordia*-plantain system, followed in descending order by taungya, new cacao, old cacao, line plantings and pure plantations. *Acacia* in association with maize, ginger, and araza performed much better than in pure plantations in a study carried out in Talamanca, Costa Rica (Kapp and Beer, 1995). However, the authors did not recommend this species for alluvial soils in this area because of its susceptibility to the root fungal disease carried by *Rosellina* spp., and rodent damage in the seedling stage. *E. deglupta* and *Cordia* plantations can be successfully established in a taungya system if associated crops and tree-crop distances are chosen adequately (Schlönvoigt and Beer, 2001). These authors tested both species with maize and cassava (*Manihot esculenta* Crantz.). The former did not affect the species growth, but the latter did. The authors also recommend growing *Cordia* in a tree-crop distance of at least 1 m, whereas *Eucalyptus* was hardly affected by the association even at tree-crop distance of 40 cm.

Improved Fallows

The use of improved fallows has been proposed as a management alternative to shifting cultivation in the tropics (Montagnini and Mendelsohn, 1997). Improved fallows are described as the woody species planted and left to grow during the fallow phase of an agricultural field (Nair, 1990). In Coto Brus, Puntarenas, Costa Rica, a fallow enrichment trial was conducted to test the effects of five leguminous nitrogen-fixing trees on the yield of bean (*Phaseolus vulgaris* L.) (Kettler, 1997). The treatments were: (1) control: "frijol tapado" (covered beans), this is a procedure that consists of broadcasting bean seeds into weeds, and then covering them by cutting and chopping the surrounding weeds with a machete, and species fallow enrichment with (2) *Erythrina poeppigiana*,

(3) *Calliandra calothyrsus* Meissner., (4) *Gliricidia sepium*, (5) *Inga edulis*, (5) *Inga/Erythrina*, (6) *Inga/Calliandra*, and (7) *Inga/Gliricidia*. Both treatments with *Erythrina* presented high mortality and they were taken out of the analysis. All the other enrichment fallow treatments showed higher yields than the "frijol tapado" control. The control treatment for the first year of production presented a yield of 1,186 kg ha^{-1}, while the highest yield was 1,674 kg ha^{-1} for the mixed of *Inga/Gliricidia*. However, there were not statistically significant differences in yield among all the evaluated treatments.

Fallow enrichment with other species such as *Vochysia ferruginea* and *Hyeronima alchorneoides*, can be an economically viable and sustainable activity in Costa Rica (Montagnini and Mendelsohn, 1997). These authors found that by managing the forest fallows with these timber species, farmers can increase the present value of swidden agriculture from the traditional levels of US $ 0.5-1.7 to US $ 5,000-12,000 per hectare.

Silvopastoral Systems

Silvopastoral systems are the association of pastures with trees and/or animals. They can be subdivided in (a) trees on rangeland or pastures (i.e., multi-purpose trees scattered or arranged in a pasture), (b) protein bank (i.e., leguminous fodder trees), and (c) plantation crops with pastures and animals (Nair, 1990). In Costa Rica, more than 90% of the cattle farms have trees scattered throughout the pastures to provide shade for cattle. These trees also provide timber (Souza de Abreu et al., 2000), fruits for birds, fence posts, protection of the grass against desiccation in the dry season, wind protection, firewood, and other products (Harvey et al., 1998; Harvey and Haber, 1999). In Costa Rica, over 75% of the farms have living fences to separate the pastures (Souza de Abreu et al., 2000). Most of the trees and shrubs introduce themselves into the landscape matrix by natural regeneration, such as *Cordia alliodora*, *Cedrela odorata*, *Terminalia oblonga* and *Pentaclethra macroloba* (Willd.) Kuntze (Harvey et al., 1998; Harvey and Haber, 1999; Souza de Abreu et al., 2000). However, there has been an increase in the practice of planting trees for obtaining extra income and to rehabilitate degraded pastures (Budowski and Russo, 1997; Dagang and Nair, 2003; Montagnini et al., 2003). In areas where pastures are degraded by over-exploitation, silvopastoral systems in which cattle and pastures are combined with trees, shrubs and other crops may have the greatest probability of adoption by cattle farmers (Montagnini et al., 2003).

Research on timber species for silvopastoral systems should concentrate on selecting high-value commercial species that can permit or favor the growth of pastures and tolerate the presence of cattle (Montagnini et al., 2003). For instance, *Eucalyptus deglupta* and *Acacia mangium* have been shown to grow well in pastures and they do not affect negatively pasture productivity in the Atlantic lowlands of Costa Rica (Andrade et al., 2000). Montagnini et al. (2003) suggest the following species for silvopastoral systems in degraded areas of the Atlantic lowlands: *Vochysia guatemalensis*, *Calophyllum brasiliense*, *Terminalia amazonia*, *Virola koschnyi*, *Dipteryx panamensis*, *Hyeronima alchorneoides* and *Vochysia ferruginea*. The use of the foliage of species such as *Erythrina poeppigiana* and *Gliricidia sepium*, growing in silvopastoral systems for protein banks, is a protein supplement to the basic diet of cattle for milk production in Costa Rica (Camero, 1995; Camero et al., 2001). They found that milk yields did not differ ($P > 0.05$) when cows received either *E. poeppigiana* or *G. sepium* foliages as protein supplements (7.3 and 7.4 kg milk/cow/day, respectively), but they were superior ($P < 0.05$) to urea supplementation (6.7 kg milk/cow/day). The use of foliage of these species is more economical than the traditional protein supplements (Camero, 1995).

Multipurpose Trees on Crop Lands

Multi-purpose trees on crop lands are either scattered haphazardly or according to some systematic pattern on bunds, terraces, or plotfield boundaries (Nair, 1990). Agroforestry systems including multipurpose trees on crop lands with *Terminalia amazonia*, tomato (*Lycopersicon esculentum* Mill.) and vanilla (*Vanilla planifolia* L.) have been established in Costa Rica (Table 6). The fastest growth occurred when *T. amazonia* was interplanted with legume trees, especially *Inga edulis* in southwestern Costa Rica (Nichols et al., 2001). Plots with *I. edulis* interplanted with *T. amazonia* closed canopy earlier, saved efforts in hand-weeding and provided large amounts of litter biomass and edible fruit pods. Legume trees that grow well in degraded and acid soils in Costa Rica are relevant to the development of both forestry and agroforestry systems (Benites, 1990; Budowski and Russo, 1997; Tilki and Fisher, 1998).

Plots of *Erythrina lanceolata* Stand. intercropped with vanilla have performed well (Berninger and Salas, 2003). It can be used as a support tree and shade tree. It provided a higher biomass production and N yield in a trial in southwestern Costa Rica. Tomato production was studied using live stakes of *Erythrina poeppigiana* and *Gliricidia sepium* in comparison to the traditional method, which utilizes non-living stakes (Chesney et al., 2000). They obtained fruit yields of 332, 289, and 240 g plant^{-1} using two-year old stakes of *Erythrina*, *Gliricidia*, and the traditional method, respectively. Nutrient concentration in fruit dry matter was also higher in the presence of living stakes (Chesney et al., 2000).

Alley Cropping and Multi-Purpose Woody Hedgerows Systems

In alley cropping, arable annual crops are grown between hedgerows of planted shrubs and trees–preferably leguminous species–which are periodically pruned to prevent shading the companion crop(s), and which can be used for producing mulch and green manure to improve soil fertility and produce high-quality fodder (F. Montagnini, pers. com. 11 February 2004). Multi-purpose woody hedgerows can be used as woody hedges for fodder, mulch, green manure, and soil conservation (Nair, 1990). Production of maize and beans using mulch and alley cropping has been tested in Costa Rica (Table 6). To evaluate the yields of bean in alley cropping, a trial was set up using six treatments: (1) monoculture of bean; (2) bean in an alley cropping of *Erythrina berteroana* Urb. at 0.5 m × 4 m; (3) bean in an alley cropping of *E. berteroana* at 0.6 m × 6 m; (4) bean with a mulch of *Mucuna deeringiana* (Bort.) Merr.; (5) bean with a mulch of *E. berteroana*; and (6) bean with a daily application of manure. The bean yields (kg ha^{-1}) obtained for the first harvest cycle were: 268, 86, 49, 275, 642, 29, and 270; for the second cycle they were 86, 177, 167, 98, 376, 264, and 198, respectively (Tardieu et al., 2000). The best option for farmers may be the use of mulch from *E. berteroana* and manure to increase bean yields, if these resources are available. In another study the use of mulch from native trees increased the performance of maize seedlings, in comparison to unmulched controls (Montagnini et al., 1993; Byard et al., 1996). The species of their study were *Jacaranda copaia* (Aubl.) D. Don., *Stryphnodendron microstachyum*, *Vochysia guatemalensis*, *Callophylum brasiliense*, *Vochysia ferruginea* and *Hyeronima alchorneoides*.

Home Gardens

Home gardens are an intimate, multi-story combination of various trees and crops around homesteads (Nair, 1990). Two examples of home gardens in Costa Rica are presented in this paper (Table 6). The first one is a strategy for the establishment of home gardens in land reform projects (Melendez, 1996). A participatory program was implemented in order to help a new community to improve the biodiversity of their gardens. Melendez (1996) found that the establishment of a community nursery was the best means of increasing biodiversity. However, it is also mentioned that this methodology has to be adapted to the particular characteristics of each community such as its length of settlement, climate, availability of labor, and marketing potential of the products. As a result of the community nursery, the home gardens (0.6 ha of area in average) increased in diversity from 9 to 23 species in a period of three years. The home gardens included medicinal plants, and woody and fruit trees. The second study was the comparative analysis of men's and women's knowledge about thirteen medicinal and food species in home gardens of Nicoya, Costa Rica (Ochoa et al., 1998). The two principal conclusions of the research were that women had a greater knowledge of medicinal species than men and that there were almost no differences in the knowledge of men and women about food crops, except for bananas and plantains, where women had the greater knowledge. Women in most of the cases are more active in home garden activities and are the most interested in participating in community programs (Melendez, 1996).

CONCLUSIONS

This literature review presents some examples of agroforestry systems in Costa Rica, especially the role of trees and shrubs within the systems. This review, starting in the 1970s, shows a few examples of the agroforestry systems in this period which were typically systems in coffee and pasture associations. During this period there was not much knowledge of the use of native species, except for *Cordia alliodora*, *Alnus acuminata* and *Cedrela odorata* from natural regeneration. Over the past fifteen years, several new types of agroforestry systems have been studied. The studies cover a large variety of agroforestry systems and have provided more alternatives for the farmers. At present, more is known regarding native tree species and their uses for reforestation and agroforestry programs. Research in this area must continue because most of the trials are still in progress. The studies in this period are more concerned with the crop/pasture-trees interaction. They evaluate not only the role of trees on improving and maintaining soil proprieties, but also the competition that can affect the productivity of crops and trees. Financial and economic analysis has shown the effectiveness of the inclusion of tress in agricultural lands. The diversification of species in the systems is addressed to avoid risk and help the farmers. There is a general consensus that there is a need to continue investing in agroforestry research, especially in studying crop-tree interactions and the evaluation of environmental services that these ecosystems provide to society (Montagnini et al., 2003).

However, regardless the outcomes presented above, there is a lack of information regarding adoptability/acceptance of these new agroforestry techniques by the farmers. Researchers focus most on their projects' outcomes, forgetting to include the interest of farmers in adopting these techniques as a factor. It is not clear for most of the examples presented here, whether the new outcomes have been implemented or they are only part of experimental trials. In short, further social research has to be conducted in order to evaluate the implementation of these new techniques by local farmers and the reasons why. Agroforestry systems are means to improve farmer welfare and environmental conditions, then, we have to address our attention not only to the experimental research, but also to the human aspects.

REFERENCES

Andrade, H., Muhammad, I., Jiménez, F., Finegan, B. and D. Kass. 2000. Dinámica productiva de sistemas silvopastorales con *Acacia mangium* y *Eucalyptus deglupta* en el trópico húmedo. Agroforestería en las Américas 7:50-52.

Babbar, L.I. and D.R. Zak. 1994. Nitrogen cycling in coffee agroecosystems: Net N mineralization and nitrification in the presence and absence of shade trees. Agriculture, Ecosystems and Environment 48:107-113.

Beer, J.W., Clarkin, K.L., De las Salas, G., and N.L. Glover. 1979. A case study of traditional agroforestry practices in a wet tropical zone: The "La Suiza" Project. Paper presented at the Simposio internacional sobre las ciencias forestales y su contribución al desarrollo de la América Tropical. CONICIT-INTERCIENCIA-SCITEC. San José, Costa Rica.

Beer, J., Muschler, R., Kass, D. and E. Somarriba. 1998. Shade management in coffee and cacao plantations. Agroforestry Systems 38:139-164.

Benites, J.R. 1990. Agroforestry systems with potential for acid soils of the humid tropics of Latin America and the Caribbean. Forest Ecology and Management 36:81-101.

Berninger, F. and E. Salas. 2003. Biomass production of *Erythrina poeppigiana* as influenced by shoot-pruning intensity in Costa Rica. Agroforestry Systems 57:19-28.

Berti, G. 1999. Transformaciones recientes en la industria y la política forestal Costarricense y sus implicaciones para el desarrollo de los bosques secundarios. MSc. Thesis, CATIE, Turrialba, Costa Rica.

Budowski, G. and R. Russo. 1997. Nitrogen-fixing trees and nitrogen fixation in sustainable agriculture: Research challenges. Soil Biology and Biochemistry 29: 767-770.

Butterfield, R.P. 1995. Promoting biodiversity: Advances in evaluating native species for reforestation. Forest Ecology and Management 75:111-121.

Butterfield, R.P. 1996. Early species selection for tropical reforestation: A consideration of stability. Forest Ecology and Management 81:161-168.

Byard, R., Lewis, K.C. and F. Montagnini. 1996. Leaf litter decomposition and mulch performance from mixed and pure monospecific plantations of native species in Costa Rica. Agriculture, Ecosystems and Environment 58:145-155.

Camero, A. 1995. Experiencias desarrolladas por el CATIE en el uso del follaje de *Erythrina* sp. y *Gliricidia sepium* en la producción de carne y leche de bovinos. Agroforestería en las Américas 8:9-13.

Camero, A., Ibrahim, M. and M. Kass. 2001. Improving rumen fermentation and milk production with legume-tree fodder in the tropics. Agroforestry Systems 51:157-166.

Chesney, P.E., Schlönvoigt, A. and D. Kass. 2000. Producción de tomate con soportes vivos en Turrialba, Costa Rica. Agroforestería en las Américas 7:57-60.

Combe, J. 1979. Técnicas agroforestales para los trópicos húmedos: Conceptos y perspectivas. Paper presented at the Simposio internacional sobre las ciencias forestales y su contribución al desarrollo de la América Tropical. CONICIT-INTERCIENCIA-SCITEC. San José, Costa Rica.

Dagang, A.B.K. and P.K.R. Nair. 2003. Silvopastoral research and adoption in Central America: Recent findings and recommendations for future directions. Agroforestry Systems 59:149-155.

Díaz, W., Szott, L., Arcos, M., Arévalo, L. and J. Pérez. 1993. Análisis y evaluación económica del cultivo de pijuayo en sistemas agroforestales. In: pp. 323-346. J. Mora, L. Szott, L. Murillo, M. and V. Patiño (eds). IV Congreso internacional sobre biología, agronomía e industrialización del Pijuayo. Iquitos, Perú.

Fassbender, H.W., Beer, J., Heuveldop, J., Imbach, A., Enriquez, G. and A. Bonnemann. 1991. Ten year balances of organic matter and nutrients in agroforestry systems at CATIE, Costa Rica. Forest Ecology and Management 45: 173-183.

Ford, L.B. 1979. Estimación del rendimiento del *Cedrela odorata* L. cultivado en asocio con café. In: pp. 183-189. De las Salas, G. (Ed). Taller sistemas agroforestales en América Latina. CATIE-Universidad de las Naciones Unidas, Turrialba, Costa Rica.

Fournier, L.A. 1979. El cultivo de jaúl (*Alnus jorullensis*) en fincas de café en Costa Rica. In: pp. 163-167. De las Salas, G. (Ed). Taller sistemas agroforestales en América Latina. CATIE-Universidad de las Naciones Unidas, Turrialba, Costa Rica.

Fournier, L.A. 1981. Importancia de los sistemas agroforestales en Costa Rica. Agronomía Costarricense 5:141-147.

González, M., Martínez, H. and N. Gewald. 1979. El uso de prácticas silvopastoriles en las partes altas del Valle Central de Costa Rica. In: pp. 208-210. De las Salas, G.

(Ed). Taller sistemas agroforestales en América Latina. CATIE-Universidad de las Naciones Unidas, Turrialba, Costa Rica.

González, E.J. and R.J. Fisher. 1994. Growth of native species planted on abandoned pasture land in Costa Rica. Forest Ecology and Management 70:159-167.

Haggar, J.P., Briscoe, C.B. and R.P. Butterfield. 1998. Native species: A resource for the diversification of forestry production in the lowland humid tropics. Forest Ecology and Management 106:195-203.

Harvey, C.A., Haber, W.A., Mejias, F. and R. Solano. 1998. Remnant trees in Costa Rica pastures: Tools for conservation. Agroforestry Today 10:7-9.

Harvey, C.A. and W.A. Haber. 1999. Remnant trees and the conservation of biodiversity in Costa Rica pastures. Agroforestry Systems 44:37-68.

Kapp, G.B. and J. Beer. 1995. A comparison of agrisilvicultural systems with plantation forestry in the Atlantic Lowlands of Costa Rica. Agroforestry Systems 32:207-223.

Kapp, G.B., Beer, J. and R. Lujan. 1997. Species and site selection for timber production of farm boundaries in the humid Atlantic lowlands of Costa Rica and Panama. Agroforestry Systems 35:139-154.

Kettler, J.S. 1997. Fallow enrichment of a traditional slash/mulch system in southern Costa Rica: Comparison of biomass and crop yield. Agroforestry Systems 35:165-176.

Leal, D., Kass, D., Lok, R., Köpsell, E. and I. Muhammad. 2000. Evaluación participativa de alternativas agroforestales para la producción de palmito (*Bactris gasipaes*) en tierras de ladera del Atlántico de Costa Rica. Agroforestería en las Américas 7:14-16.

Leopold, A.C., Andrus, R., Finkeldey, A. and D. Kwoles. 2001. Attempting restoration of wet tropical forests in Costa Rica. Forest Ecology and Management 142:243-249.

Lyngbæk, A.E., Muschler, R.G. and F.L. Sinclair. 2001. Productivity and profitability of multistrata organic versus conventional coffee farms in Costa Rica. Agroforestry Systems 53:205-213.

Mehta, N.G. and Leuschner, W.A. 1997. Financial and economic analyses of agroforestry systems and a commercial timber plantation in the La Amistad Biosphere Reserve, Costa Rica. Agroforestry Systems 37:175-185.

Meléndez, L. 1996. Estrategia para el establecimiento de huertos caseros en asentamientos campesinos en el Área de Conservación de Tortuguero, Costa Rica. Agroforestería en las Américas 9:25-28.

Montagnini, F., Stijhoorn, E. and F. Sancho. 1989. Soil chemical properties and root biomass under plantations of native tree species, grass cover and secondary forest vegetation in the Atlantic lowlands of Costa Rica. Belowground Ecology Bulletin 1:6-8.

Montagnini, F., Ramstad, K. and F. Sancho. 1993. Litterfall, litter decomposition and the use of mulch of four indigenous tree species in the Atlantic lowlands of Costa Rica. Agroforestry Systems 23:39-61.

Montagnini, F. and F. Sancho. 1994a. Aboveground biomass and nutrients in young plantations of indigenous trees on infertile soils in Costa Rica: Implications for site nutrient conservation. Journal of Sustainable Forestry 1:115-139.

Montagnini, F. and F. Sancho. 1994b. Net nitrogen mineralization in soils under six indigenous tree species, an abandoned pasture and a secondary forest in the Atlantic lowlands of Costa Rica. Plant and Soil 162:117-124.

Montagnini, F., and R. Mendelsohn. 1997. Managing forest fallows: Improving the economics of swidden agriculture. Ambio 26:118-123.

Montagnini, F. 2000. Accumulation in above-ground biomass and soil storage of mineral nutrients in pure and mixed plantations in a humid tropical lowland. Forest Ecology and Management 134:257-270.

Montagnini, F., Ugalde, L. and C. Navarro. 2003. Growth characteristics of some native tree species used in silvopastoral systems in the humid lowlands of Costa Rica. Agroforestry Systems 59:163-170.

Montenegro, J. and A. Bogantes. 2001. Palmito de pejibaye (*Bactris gasipaes*) cultivado bajo diferentes densidades de laurel (*Cordia alliodora*). Agronomía Costarricense 25:73-79.

Moreira, A.L. and A. Palma. 1987. Boletín estadístico No. 2. Dirección General Forestal, San José, Costa Rica.

Muschler, R.G. and A. Bonnemann. 1997. Potentials and limitations of agroforestry for changing land-use in the tropics: Experiences from Central America. Forest Ecology and Management 91:61-73.

Muschler, R.G. 2001. Shade improves coffee quality in a sub-optimal coffee-zone of Costa Rica. Agroforestry Systems 85:131-139.

Nair, P.K.R. 1990. Classification of agroforestry systems. In: pp. 31-57. MacDicken, K. and N. Vergara (eds.). Agroforestry: Classification and management. Wiley, New York, USA.

Nichols, J.D., Rosemeyer, M.E., Carpenter, F.L. and J. Kettler. 2001. Intercropping legume trees with native timber trees rapidly restores cover to eroded tropical pastures without fertilization. Forest Ecology and Management 152:195-209.

Nygren, P. and Ramírez, C. 1995. Production and turnover of N_2 fixing nodules in relation to foliage development in periodically pruned *Erythrina poeppigiana* (Leguminosae) trees. Forest Ecology and Management 73:59-73.

Ochoa, L., Fassaert, C., Somarriba, E. and A. Schlönvoigt. 1998. Conocimiento de mujeres y hombres sobre las especies de uso medicinal y alimenticio en huertos caseros de Nicoya, Costa Rica. Agroforestería en las Américas 5:7-10.

Piotto, D., Montagnini, F., Ugalde, L. and M. Kanninen. 2003. Performance of forest plantations in small and medium-sized farms in the Atlantic lowlands of Costa Rica. Forest Ecology and Management 175:195-204.

Piotto, D., Víquez, E., Montagnini, F., Kanninen, M. 2004. Pure and mixed forest plantations with native species of the dry tropics of Costa Rica: A comparison of growth and productivity. Forest Ecology and Management 190:359-372.

Poschen, P. 1980. El jaúl con pasto, la práctica de un sistema silvopastoril en Costa Rica. Facultad de Ciencias Forestales, Universidad de Friburgo, Alemania.

Ramírez, O.A., Somarriba, E., Ludewigs, T. and P. Ferreira. 2001. Financial returns, stability and risk of cacao-plantain-timber agroforestry systems in Central America. Agroforestry Systems 51:141-154.

Rosero, P. and N. Gewald. 1979. Crecimiento de laurel (*Cordia alliodora*) en cafetales, cacaotales y potreros en la zona Atlántica de Costa Rica. In: pp. 211-214. De las Salas, G. (Ed). Taller sistemas agroforestales en América Latina. CATIE-Universidad de las Naciones Unidas, Turrialba, Costa Rica.

Sauer, J.D. 1979. Living fences in Costa Rican agriculture. Turrialba 29:255-261.

Schaller, M., Schroth, G., Beer, J. and Jiménez, F. 2003. Species and site characteristics that permit the association of fast-growing trees with crops: The case of *Eucalyptus deglupta* as coffee shade in Costa Rica. Forest Ecology and Management 175:205-215.

Schlönvoigt, A. and M. Schlönvoigt. 1999. Intercropping *Cordia alliodora* (Ruiz & Pavón) Oken plantations with *Bactris gasipaes* H.B.K. in San Carlos, Costa Rica. In: pp. 135-138. Jiménez, F. and J. Beer (comp.) International Symposium: Multistrata agroforestry systems with perennials crops, Turrialba, Costa Rica.

Schlönvoigt, A. and J. Beer. 2001. Initial growth of pioneer timber tree species in a Taungya system in the humid lowlands of Costa Rica. Agroforestry Systems 51:97-108.

Somarriba, E., Beer, J. and R.G. Muschler. 2001a. Research methods for multistrata agroforestry systems with coffee and cacao: Recommendations from two decades of research at CATIE. Agroforestry Systems 53:195-203.

Somarriba, E., Valdivieso, R., Vásquez. and Galloway, G. 2001b. Survival, growth, timber productivity and site index of *Cordia alliodora* in forestry and agroforestry systems. Agroforestry Systems 51:111-118.

Souza de Abreu, M.H., Muhammad, I. Harvey, C. and F. Jiménez. 2000. Caracterización del componente arbóreo en los sistemas ganaderos de La Fortuna de San Carlos, Costa Rica. Agroforestería en las Américas 7:53-56.

Stanley, W.G. and F. Montagnini. 1999. Biomass and nutrient accumulation in pure and mixed plantations of indigenous tree species grown on poor soils in the humid tropics of Costa Rica. Forest Ecology and Management 113:91-103.

Tardieu, R., Kass, D. and A. Oliver. 2000. Efecto de prácticas agroforestales y agrícolas sobre el rendimiento de frijol y la disponibilidad de fósforo en un andisol de Costa Rica. Agroforestería en las Américas 7:61-64.

Tavares, F.C., Beer, J., Jiménez, F., Schroth, G. and C. Fonseca. 1999. Costa Rican farmers' experience with the introduction of timber trees in their coffee plantations. In: pp. 268-271. Jiménez, F. and J. Beer (comp.) International Symposium: Multistrata agroforestry systems with perennials crops, Turrialba, Costa Rica.

Tilki, F. and R.F. Fisher. 1998. Tropical leguminous species for acid soils: Studies on plant form and growth in Costa Rica. Forest Ecology and Management 108:175-192.

Tornquist, C.G., Hons, F.M., Feagley, S.E., and J. Haggar. 1999. Agroforestry system effects on soil characteristics of the Sarapiquí region of Costa Rica. Agriculture, Ecosystems and Environment 73:19-28.

Young, A. 1997. Agroforestry for soil management. 2nd Edition. C.A.B. International, Wallingford, UK.

Index